Video and TV Projects

The Maplin series

This book is part of an exciting series developed by Butterworth-Heinemann and Maplin Electronics Plc. Books in the series are practical guides which offer electronic constructors and students clear introductions to key topics. Each book is written and compiled by a leading electronics author.

Other books published in the Maplin series include:

Computer Interfacing	Graham Dixey	0 7506 2123 0
Logic Design	Mike Wharton	0 7506 2122 2
Music Projects	R A Penfold	0 7506 2119 2
Starting Electronics	Keith Brindley	0 7506 2053 6
Audio IC Projects	Maplin	0 7506 2121 4
Integrated Circuit Projects	Maplin	0 7506 2578 3
Auto Electronics Projects	Maplin	0 7506 2296 2
Test Gear & Measurement	Danny Stewart	0 7506 2601 1
Home Security Projects	Maplin	0 7506 2603 8
The Maplin Approach to Professional Audio	T.A. Wilkinson	0 7506 2120 6

Video and TV Projects

Newnes
An imprint of Butterworth-Heinemann Ltd
Linacre House, Jordan Hill, Oxford OX2 8DP

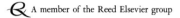 A member of the Reed Elsevier group

OXFORD   LONDON   BOSTON
MUNICH   NEW DELHI   SINGAPORE   SYDNEY
TOKYO   TORONTO   WELLINGTON

British Library Cataloguing in Publication Data
A catalogue record for this book is available from the
British Library
ISBN 0 7506 2297 0

Library of Congress Cataloguing in Publication Data
A catalogue record for this book is available from the
Library of Congress

 Edited by Co-publications, Loughborough

 Typeset and produced by Sylvester North, Sunderland

all part of The Sylvester Press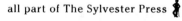

Printed in Great Britain

Contents

Preface

This book is a collection of project articles previously published in *Electronics — The Maplin Magazine*. They were chosen for publication in book form not only because they are were so popular with readers in their original magazine appearances but also because they are so relevant in the field of modern video and television — a subject area in which it is evermore difficult to find information of a technical, knowledgeable, yet understandable nature. This book, we think, is exactly that.

This is just one of the Maplin series of books published by Newnes books covering all aspects of computing and electronics. Others in the series are available from all good bookshops.

Maplin Electronics Plc publishes the monthly electronics magazine *Electronics*; it is the ideal choice for anyone who wants to keep up with the world of electronics, computing, science and technology. Practical electronic projects are included with all parts readily available.

Maplin Electronics Plc also supplies a wide range of other electronic components and associated products to private individuals and trade customers. Telephone (01702) 552911 or write to Maplin Electronics, PO Box 3, Rayleigh, Essex SS6 8LR, for further details of product catalogue and locations of regional stores.

1 Audio and video modulator

When using certain video equipment, an ordinary television receiver can not be connected directly to the video signal. Some TV sets do have a direct video input socket (SCART), but most domestic sets only have an aerial input for reception of UHF TV stations. To solve this problem, a UHF modulator is required, which superimposes video and audio signals on to a high frequency carrier wave. To simplify the construction and alignment of the project a pre-tuned modulator module (UM1286) has been employed in the design. From the composite video and mono sound signals the modulator produces an RF output suitable for connection to the aerial input of a UK UHF TV set. The carrier frequency is chosen to fall on an unused television channel (channel 36). The UM1286 has an integral 6 MHz RF oscillator for the sound subcarrier signal and a wide linear video bandwidth to cater for the chrominance sub-carrier. A few of the many possible configurations of audio/video equipment connections to and from the finished boxed unit are shown in Figures 1.1(a), (b), and (c).

Circuit description

A block diagram of the project is shown in Figure 1.2. This should assist you when following the circuit description or fault finding in the completed unit. Figure 1.3 shows the complete circuit.

The d.c. power is applied to PL2, positive voltage to pin 1 and negative to pin 4 (0 V). This supply must be within the range 8 V to 16 V and have the correct polarity, otherwise damage will occur to the semiconductors and polarised components. To prevent a supply of opposite polarity being applied to the circuit, diode D2 has to have the positive supply voltage applied to its anode before the d.c. power can pass to the rest of the circuit.

Resistor R9 and capacitor C6 provide the main decoupling for the +V supply rail, with capacitor C5 giving additional high frequency decoupling. Further supply decoupling is provided by R15, C15 and C16 in the +1 V audio supply rail. The red LED power on indicator, LD1 has its anode connected to pin 2 of PL2 and its cathode to pin 3. Resistor R8 provides current limiting to the LED thus restricting the drain to only a few mA.

Audio and video modulator

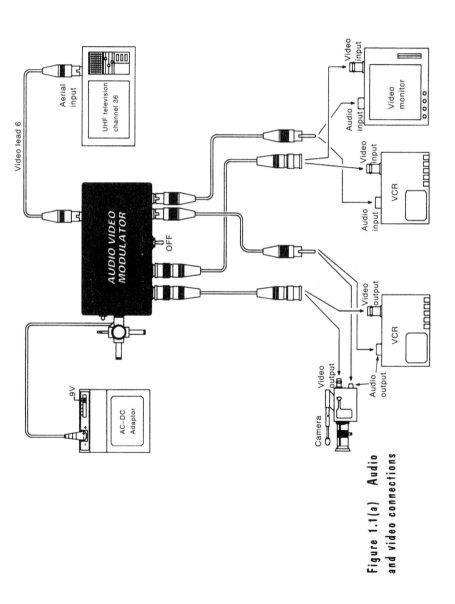

Figure 1.1(a) Audio and video connections

3

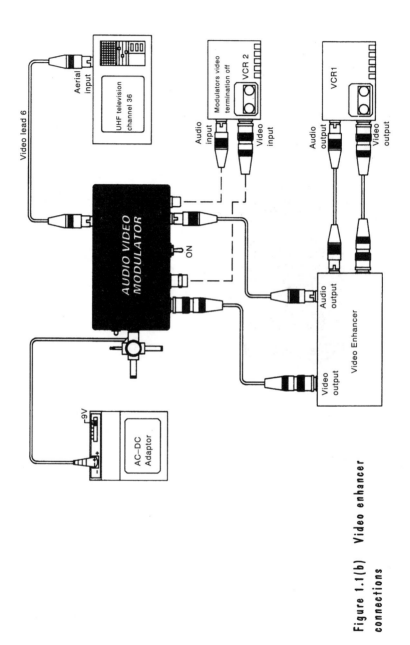

Figure 1.1(b) Video enhancer connections

Audio and video modulator

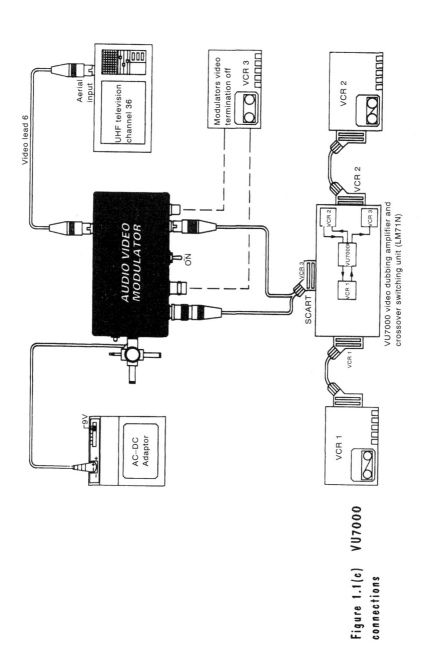

Figure 1.1(c) VU7000 connections

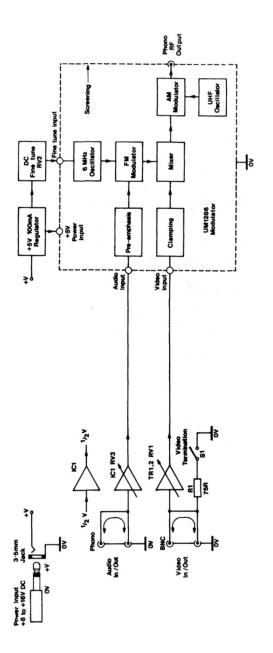

Figure 1.2 Block diagram

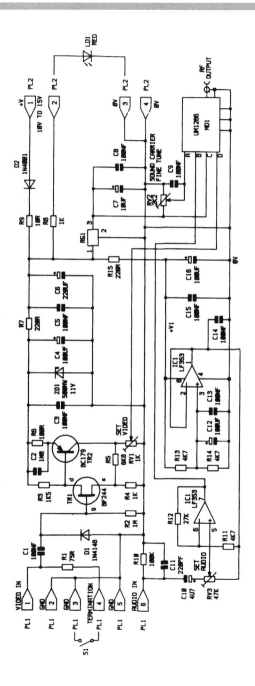

Figure 1.3   Circuit diagram

Video and TV projects

The circuit incorporates two voltage regulators; an 11 V
zener diode, ZD1 and a +5 V regulator integrated circuit,
RG1. Zener diode ZD1, in conjunction with resistor R7,
limits the voltage supply to the video buffer, this rail is
decoupled by capacitors C3 and C4. The +5 V output from
regulator RG1 is used to power the UM1286, MD1 (pin C)
and also provides a voltage reference to RV2 — the sound
sub-carrier oscillator fine tuning control. Capacitors C7
and C8 are used to decouple this +5 V supply, with capacitor C9 decoupling the fine tune input of MD1 (pin
A).

For the audio circuit to function correctly a half +V1 supply reference is necessary. This is provided by half of
IC1. The voltage reference applied to the input of this
op-amp is derived from the two resistors R13 and R14
which form a potential divider. The op-amp is merely
used as a zero gain buffer to provide a low impedance
half supply, its input being decoupled by capacitor C12
and C13, and its output by capacitor C14. The other half
of this IC is used as an audio amplifier which drives the
sound input of the UM1286, MD1 (pin B). Resistors R11
and R12 are used to set the gain of the op-amp with RV3
adjusting the level of audio signal applied to its input.
The incoming signal from pin 6 of PL1 is a.c. coupled to
this control via C10 with the input being loaded by R10
and C11. Pin 5 of PL1 is used as the ground or screen
connection for the audio line.

The video signal is applied to pin 1 of PL1 and its ground
is connected to pin 2. The input impedance of the video
amp is approximately 1 MΩ. However, this input can be
reduced to 75 Ω by operating switch S1 on pins 3 and 4
of PL1. When the switch is closed 75 Ω resistor R1 is

placed across the video input, this is known as a termination load.

The video signals are a.c. coupled via C1 into the gate of the field effect transistor (FET) TR1, with diode D1 and resistor R2 used to maintain the correct bias level. Transistors TR1 and TR2 form a broad band buffer amplifier, with its gain set by the value of the negative feed back resistor R5. Resistor R4 is used as the source load for TR1 and the preset RV1 as the collector output load in TR2. The d.c. bias for TR2 is derived from R3 and TR1, and a small amount of frequency compensation is provided by C2 and R6.

The video output from the amplifier is tapped off by the wiper of RV1 and is fed to the video input pin (D) of the UM1286 modulator MD1. Inside MD1 the audio signal is converted into a 6 MHz FM modulated sub-carrier. It is then mixed with the video signal and fed to the AM modulator where the UHF carrier is combined to produce the final modulated RF output.

PCB assembly

Take note — Take note —Take note — Take note

The PCB is a single-sided fibre glass type, chosen for maximum reliability and stability. However, removal of a misplaced component is quite difficult so please double-check each component type, value and its polarity where appropriate, before soldering!

Video and TV projects

The PCB has a printed legend to assist you in correctly positioning each item (see Figure 1.4). The sequence in which the components are fitted is not critical. However, the following instructions will be of use in making these tasks as straightforward as possible. It is usually easier

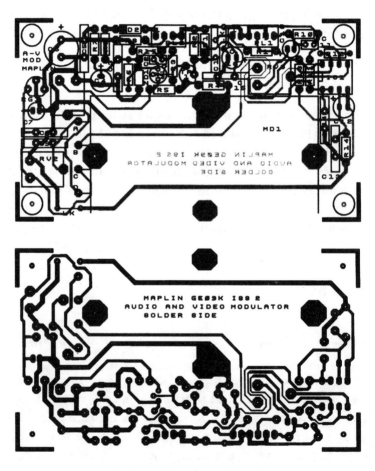

Figure 1.4   Track and layout of the PCB

to start with the smaller components, such as the resistors. Next mount the ceramic and electrolytic capacitors. The polarity for the electrolytic capacitors is shown by a plus sign (+) matching that on the PCB legend. However, on some capacitors the polarity is designated by a negative symbol (–), in which case the lead nearest this symbol goes away from the positive sign on the legend. All the diodes have a band at one end. Be sure to position them according to the legend, where the appropriate markings are shown.

Next install the two transistors and the voltage regulator, matching each case to its outline on the legend. When fitting the eight pin IC socket ensure that you match the notch with the block on the board. Install IC1 making certain that all the pins go into the socket and the pin one marker is at the notched end. Next install the three preset resistors RV1, 2, 3 and set them all to their half way positions. When fitting the *Minicon* connectors ensure that the locking tags are facing inwards. Using component lead off-cuts fit the wire links at the two positions marked LK on the PCB. Finally, mount the UM1286 modulator MD1, making certain that all four wire connections are in their correct positions (A, B, C and D). To secure MD1 to the PCB simply twist the four fixing tags through 90 degrees, and using a fair amount of solder, heat in to place.

This completes the assembly of the PCB and you should now check your work very carefully making sure that all the solder joints are sound. It is also very important that the solder side of the circuit board does not have any trimmed component leads standing proud of the soldered track, as this may result in a short circuit when the unit is fitted into its metal die-cast box.

Wiring

If you buy the hardware kit from Maplin it should contain a one metre length of hook-up wire. Once the PCB
assembly has been fitted inside its die-cast box it becomes difficult to fault find, and for this reason it is
advisable to make temporary connections to the PCB and
chassis sockets, as shown in Figure 1.5. At this stage the
wires can be longer than required as they are cut to size
during final assembly. The starting point of each wire is
taken from a terminal in the Minicon connector PL1 or
PL2. The terminals must be crimped then soldered to
each wire before it is inserted into the Minicon housing,
a locking tag on the terminal will ensure that it stays
securely in place.

Testing

All the tests can be made with a minimum of equipment.
You will need a multimeter, a UHF TV set and an audio/
video source. To power the unit you will require a +8 V
to +16 V d.c. supply, and an unregulated a.c. adaptor type
(XX09K from Maplin) set to its +9 V output is adequate.
The readings were taken on the prototype using a digital multimeter and some of the readings obtained may
vary slightly depending on the type of meter you use.

Carefully lay out the PCB assembly on a non-conductive
surface, such as a piece of dry paper or plastic. Position
the chassis mounting components so they are clear of
the circuit board and make sure the wires are as shown
in Figure 1.5. The d.c. input jack socket is a type commonly used on Japanese radio equipment, where the

Audio and video modulator

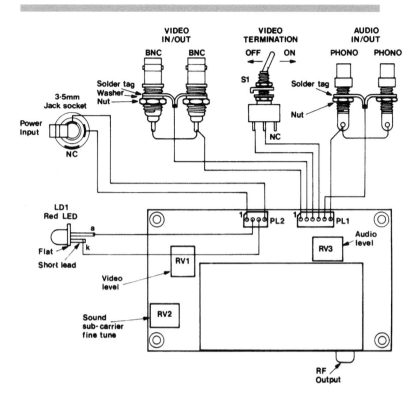

Figure 1.5   Wiring

centre pin is the positive connection and the negative contact is the threaded body. The first test is to measure the resistance at this socket. With your multimeter set to read ohms, connect its red positive test lead to the terminal with the wire going to pin 1 of PL2 and connect the black negative lead to the other terminal. You should get a reading of approximately 1.8 kΩ and when the test leads are reversed, a much higher reading in excess of 20 MΩ should be present. These readings are due to diode D2, the component which protects the rest of the circuit from reverse polarity damage.

13

Video and TV projects

In the following tests it will be assumed that the power supply used is the unregulated a.c. adaptor set to its +9 V output. Select a suitable range on your meter that will accommodate a 300 mA d.c. current reading and place it in the positive power line from the jack socket. Connect the 3.5 mm jack plug of the mains adaptor to the power input, then plug the adaptor into the a.c. mains supply. The power indicator LD1 should light up, with a current reading of approximately 40 mA being observed. Unplug the adaptor from the mains, then remove the test meter and reconnect the positive line to the jack socket.

Now set your multimeter to read d.c. volts. All voltages are positive with respect to ground, so connect your negative lead to a convenient ground point on the unit. When the modulator is powered up, voltages present on the PCB should approximately match the following:

Pin 1 of PL2	=	+14.5 V d.c.
Pin 2 of PL2	=	+2 V d.c.
Pin C of MD1	=	+5 V d.c.
Pin 8 of IC1	=	+12 V d.c.
Pin 1 of IC1	=	+6 V d.c.
Cathode of ZD1	=	+11 V d.c.

This completes the d.c. testing of the audio and video modulator, now remove your multimeter from the unit.

Next connect a phono to coax lead (Maplin code FV90X) from the RF output of the modulator to the aerial input of a UHF television, see Figure 1.1(a). Using a spare channel selector tune to approximately 36, where you should find a blank screen and a silent sound track. Connect an audio/video signal to the in/out of the modulator, if no other video connection is made to the unit then the ter-

14

mination switch must be on. To set the audio level, adjust RV3 until the sound level is the same as an off air transmission (BBC, ITV, CH4). Next set the video level so that peak whites don't flare out and produce excessive buzzing on the sound channel. If this buzzing sound persists you can try tuning it out using RV2; the sound sub-carrier fine tune. The final setting of video level is up to personal choice.

Do not make any attempt to adjust the presets inside the UM1286 modulator, as these are factory set using sophisticated test equipment.

Final assembly

The unit is designed to fit in to a die-cast metal box type M5004 (Maplin code LH71N) which is also available ready drilled (Maplin code YT64U). However, if you wish to make up your own box, drilling details are given in Figure 1.6.

Next remove all the chassis mounting components from the wiring and disconnect the Minicon plugs from the PCB assembly. The PCB will only just fit inside the box so the following procedure must be used:

●   remove the metal screening lid of the UM1286 modulator,

●   position the PCB at an angle to the box so that the phono socket of the modulator passes through the hole in the side,

Video and TV projects

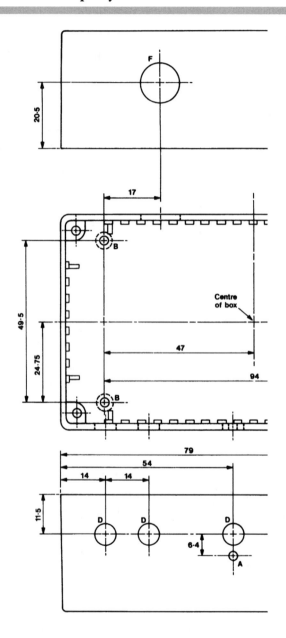

Figure 1.6   Box drilling

Audio and video modulator

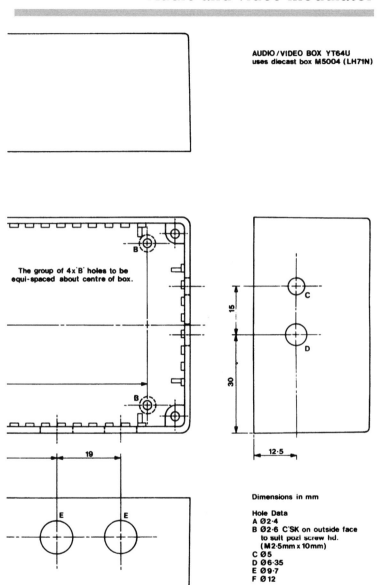

AUDIO/VIDEO BOX YT64U
uses diecast box M5004 (LH71N)

The group of 4 x B holes to be
equi-spaced about centre of box.

Dimensions in mm

Hole Data
A Ø2·4
B Ø2·6 C'SK on outside face
   to suit pozi screw hd.
   (M2·5mm x 10mm)
C Ø5
D Ø6·35
E Ø9·7
F Ø12

Third angle projection

Figure 1.6   Continued

Video and TV projects

- carefully position the PCB and secure in to place using the M2.5 hardware,

- refit the screening of the modulator.

Install the BNC and phono sockets ensuring that all are tightly secured with their solder tags facing each other, as shown in Figure 1.5. Next fit the video termination switch S1 and the power input jack socket. The red LED power indicator LD1 is held in position by a 3 mm panel mounting clip which is simply pushed in to place.

This completes the assembly of the unit. Now refit the Minicon plugs and rewire the chassis mounted components, as shown in Figure 1.5. Before fitting the custom made stick-on top panel (stock code JL74R) test out the unit to ensure that all is well. Finally fit the lid of the box using the screws provided and stick on the four small rubber feet. The unit is now ready for use.

Using the modulator

The audio/video modulator has been designed to be tolerant to varying supply voltages and differing inter-connecting lead lengths. The following information should assist you in setting up your system.

A.C.–D.C. adaptor model XX09K

Minimum voltage setting = 6 V.
Normal voltage setting = 9 V.
Maximum voltage setting = 12 V.

Rev change = plus sign (+) on d.c. output plug to + on adaptor.

Phono to coax lead length

Minimum length = as short as you like.
Normal length = 1.2 metres (video lead 6).
Maximum length = 10 metres (good quality low-loss co-axial cable).

Phono audio lead length

Minimum length = as short as you like.
Normal length = 1.5 metres (video lead 4) or 1.2 metres (plugpak 279).
Maximum length = 4 metres (good quality low noise cable).

BNC video lead length

Minimum length = as short as you like.
Normal length = 1.5 to 1.8 metres (video lead 1, 3, or 5).
Maximum length = 4 metres (good quality low-loss cable).

Unterminated video input to modulator

Video termination switch = *on* (see Figures 1(b) and 1(c)).

Terminated video input to/from modulator

Video termination switch = *off* (see Figure 1.1(a)).

Video and TV projects

Power supply voltage:	8 V to 16 V d.c.
Supply current at 8 V:	26 mA
12 V:	32 mA
16 V:	48 mA
Audio input level:	1 V peak-to-peak
Audio input impedance:	30 kΩ
Vide input level:	1 V peak-to-peak
Video input impedance:	1 MΩ (no termination)
	75 Ω (terminated)
RF TV output:	Channel 36 (591.5 MHz)
Sound sub-carrier:	6 MHz
Video bandwidth:	8 MHz
Output socket:	Phono

Table 1.1 Specification of prototype

Hardware parts list

Miscellaneous

case drilled	1		(YT64U)
stick-on feet small	1	pkt	(FE32K)
stick-on top panel	1		(JL74R)
3 mm LED clip	1		(YY39N)
M2.5 x 10 mm pozi screw	1	pkt	(JC68Y)
M2.5 steel nut	1	pkt	(JD62S)
M2.5 isoshake	1	pkt	(BF45Y)
3.5 mm jack socket	1		(HF82D)
chassis phono socket	2		(YW06G)
50 Ω BNC round socket	2		(HH18U)
hook-up wire blue	1	pkt	(BL01B)

Audio and video modulator parts list

Resistors — All 0.6 W 1% metal film

R1	75 Ω	1	(M75R)
R2	1 M	1	(M1M)
R3	1k5	1	(M1K5)
R4,8	1 k	2	(M1K)
R5	6k8	1	(M6K8)
R6	100 Ω	1	(M100R)
R7,15	220 Ω	2	(M220R)
R9	10 Ω	1	(M10R)
R10	100 k	1	(M100K)
R11,13,14	4k7	3	(M4K7)
R12	27 k	1	(M27K)
RV1	1 k hor encl preset	1	(UH00A)
RV2	2k2 hor encl preset	1	(UH01B)
RV3	47 k hor encl preset	1	(UH05F)

Capacitors

C1,3,5,8,9, 13,14,15	100 nF minidisc	8	(YR75S)

Video and TV projects

C2	1n8F ceramic	1	(WX71N)
C4,12,16	100 µF 16 V minelect	3	(RA55K)
C6	220 µF 16 V PC electrolytic	1	(FF13P)
C7	10 µF 16 V minelect	1	(YY34M)
C10	4µ7F 35 V minelect	1	(YY33L)
C11	220 pF ceramic	1	(WX60Q)

Semiconductors

D1	1N4148	1	(QL80B)
D2	1N4001	1	(QL73Q)
ZD1	BZY88C11	1	(QH15R)
LD1	mini LED red	1	(WL32K)
RG1	LM78L05ACZ	1	(QL26D)
TR1	BF244A	1	(QF16S)
TR2	BC179	1	(QB54J)
IC1	LF353	1	(WQ31J)

Miscellaneous

MD1	UM1286 modulator	1	(BK66W)
PL1	6-way minicon latch plug	1	(YW12N)
PL2	4-way minicon latch plug	1	(YW11M)
S1	sub-min toggle A	1	(FH00A)
	PC board	1	(GE09K)
	6-way minicon latch hsg	1	(BH65V)
	4-way minicon latch hsg	1	(HB58N)
	minicon terminal	1 pkt	(YW25C)
	8-pin DIL socket	1	(BL17T)
	constructors' guide	1	(XH79L)

Optional

	300 mA a.c. adaptor unreg	1	(XX09K)
	preset trimmer	1	(BR49D)
	video lead 6	1	(FV90X)

All of the above items (except Optional) are available as a kit (LM78K)

2 Audio/video generator

When setting up video equipment a stable test pattern is required for evaluating picture contrast, convergence and distortion. In this project, a single integrated circuit is used to generate all the waveforms necessary to produce greyscale, crosshatch, dot, vertical and horizontal lines. The high accuracy line and field timing is derived from a quartz crystal which provides a very stable frequency reference. The output from the generator is black and white composite video at 1 V peak-to-peak. However, not all TV sets have a direct video input socket, so a UHF modulator has been incorporated to provide an RF output on channel 36. This signal includes a 6 MHz sub-carrier for the sound channel which is modulated by a 1 kHz tone. The unit can also receive external audio and video signals from a wide range of equipment.

Video and TV projects

Circuit description

In addition to the circuit shown in Figure 2.1, a block diagram is given in Figure 2.2. This should assist you when following the circuit description or fault finding in the completed unit.

The d.c. power is applied to 2.5 mm PCB mounted socket SK1, with positive voltage on the centre pin and negative to its side terminal. This supply must be within the range of 10 V to 14 V and have the correct polarity. To prevent reverse polarity damage to the semiconductors and polarised components, diode D2 has to have the positive supply applied to its anode before the power can pass to the rest of the circuit.

The main supply rail decoupling is provided by two 1000 μF capacitors C1 and C7. Additional high frequency decoupling is incorporated by using 100 nF ceramic capacitors at regular intervals throughout the remainder of the circuit.

The video pattern generator and RF modulator circuits require a +5 V stabilised supply. This voltage is obtained by using a regulator, RG1, with capacitors C2 to C5 and C8, C9 providing supply decoupling. The red LED, LD1, is used as a power on indicator which is fed from the +5 V rail via R3.

The audio oscillator circuit IC1 is a dual op-amp with one half providing a low impedance half supply reference, while the other half is used as a Wien bridge sine-wave oscillator. Its operating frequency of 1 kHz is set by the close tolerance capacitors C13, 14 and resistors R8, 9, 11, 12. Its output of 400 mV r.m.s. is fed via R13 and C17 to P3 providing a direct oscillator output at SK3. The output is also fed via R14 and C16 to the audio source switch S1a, b. When this switch is in the out position the internal (INT) oscillator is selected. However, when pushed in external (EXT) audio signals connected to SK2 are selected. The level of both signals is controlled by RV1 feeding the input of IC2. This IC amplifies the audio signals producing an output of up to 1.7 V r.m.s. which is fed via capacitor C20 to SK4 and to terminal B of the UHF modulator MD1. The +5 V power for MD1 is applied to terminal C and a d.c. bias control, RV6, on terminal A provides fine tuning of the 6 MHz sound subcarrier.

The video patterns are generated by IC3 (shown in Figure 2.3), a ZNA234E. This device makes available all the waveforms necessary to produce grey-scale, crosshatch, dot, vertical and horizontal lines. The high accuracy timing is generated by a 2.5 MHz crystal oscillator with XT1 on pins 8 and 9 controlling its frequency. To set the width of the vertical lines, components RV3 and C25 are used to control the pulse width timing on pin 10. The grey-scale output on pin 5 is produced by a D-to-A converter from the horizontal counter. This converter is effectively a switched current sink providing 8 equal steps of approximately 60 μA/step. With the pull-up resistors R22 and RV2, 8 voltage steps are produced. Adjusting RV2 affects the grey-scale spread from black to peak white. This output requires a buffer stage, formed by transistor TR1, before it can be fed to S2. All the other video outputs have a fixed value pull up resistor and go

Video and TV projects

Page 27 →

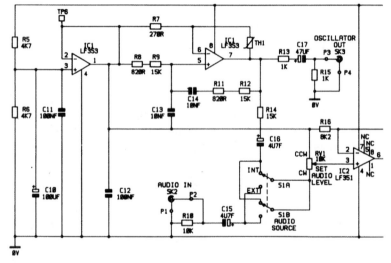

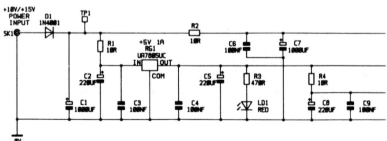

Figure 2.1   Circuit diagram

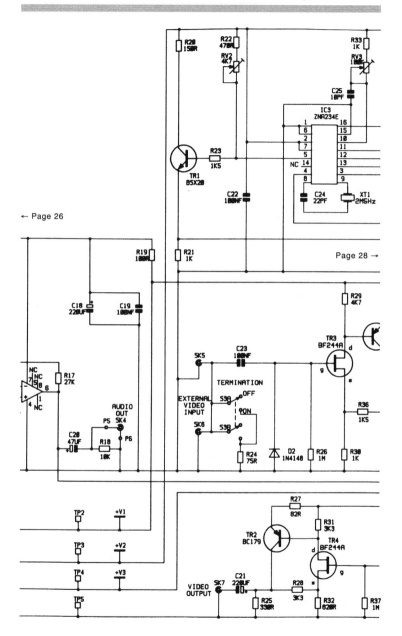

← Page 26

Page 28 →

Figure 2.1   Continued

Video and TV projects

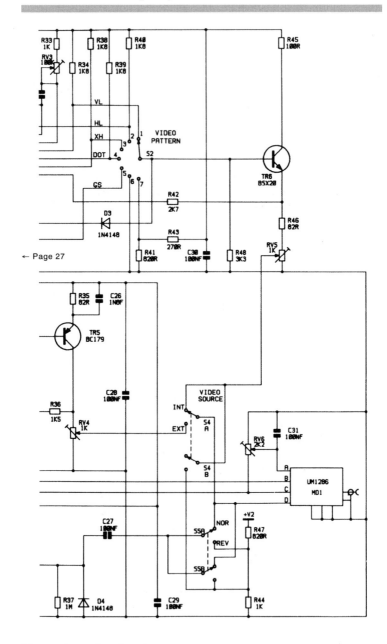

← Page 27

Figure 2.1   Continued

directly to S2 allowing selection of the pattern fed to TR6. The sync from pin 3 and blanking pulses from pin 4 are mixed to produce a composite video signal at the emitter of TR6. Variable resistor RV5 sets the level of this signal before going to the video source switch S4a and b.

With S4 in its *out* position the internal (INT) video patterns are selected. However, when pushed in any external (EXT) video signals are routed to terminal D of the RF modulator MD1 and the monitor switch S5. The external video is first processed by an amplifier with an input impedance of approximately 1 MΩ, but can be reduced to 75 Ω by operating S3. When this switch is closed a 75 Ω resistor, R24, is placed across the external video input; this is known as a termination load. The video signals are a.c. coupled via capacitor C23 into the gate of the field effect transistor (FET) TR3, with diode D1 and resistor R2 used to maintain the correct bias level. The gain of the amplifier is set by the value of the negative feed-back resistor R36. Resistor R30 is used as the source load for TR3 and the preset RV4 as the collector output load in TR5.

The video output stage formed by transistors TR2 and TR4 functions in a similar manner as the external amplifier, but receives its input from S5. This switch is used to provide the same video information at SK7 as is being modulated by MD1, or when S4 and 5 are both pushed in the external video is modulated and the test patterns reappear at SK7. This has the effect of reversing (REV) the normal (NOR) condition of the video source switching to SK7.

Video and TV projects

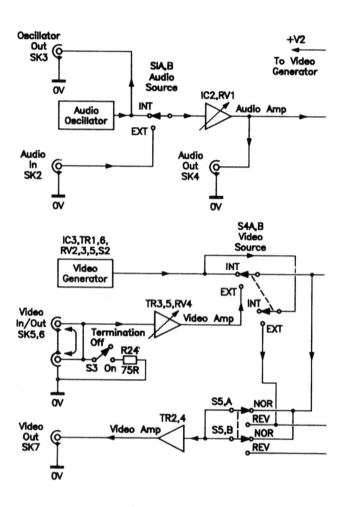

Figure 2.2   Block diagram

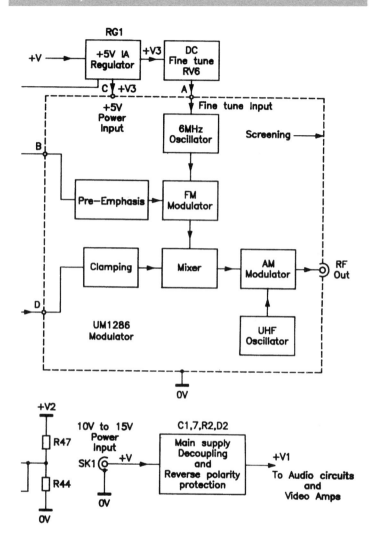

Figure 2.2 Continued

Video and TV projects

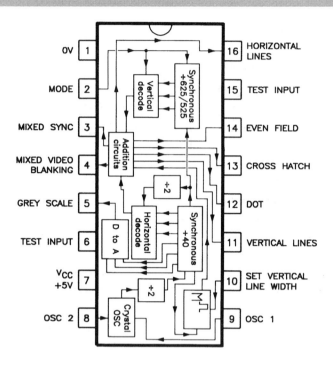

Figure 2.3 ZNA234E system diagram

PCB assembly

The PCB has a printed legend to help you correctly locate each item, shown in Figure 2.4. The sequence in which the components are fitted is not critical, however, the following instructions will be of use in making these tasks as straightforward as possible. It is usually easier to start with smaller components, such as the resistors. The 10 Ω, 1 W wirewound resistor R1, should be mounted approximately 2 mm above the surface of the PCB to allow for even heat dissipation and prevent heat damage to the PCB surface.

Next mount the ceramic, 1% polystyrene and electrolytic capacitors. The polarity for the electrolytic capacitors is shown by a plus sign (+) matching that on the PCB legend. However, on the actual body of most capacitors the polarity is designated by a negative symbol (–), in which case the lead nearest this symbol goes in the hole opposite to that adjacent to the positive sign on the legend. All the silicon diodes have a band at one end. Be sure to position them according to the legend, where the appropriate markings are shown.

Install all the transistors, matching each case to its outline. The voltage regulator, RG1, is mounted directly onto the vaned heatsink and no mica washer or heat transfer compound is required. This assembly is secured to the PCB using the M3 hardware. When fitting the IC sockets ensure that you install the appropriate one at each position, matching the notch in the end of the socket with the white block on the legend. *Do not* install the ICs until they are called for during the testing procedure!

After installing the five preset resistors RV2 to RV6 set them to their half way position. When fitting the crystal

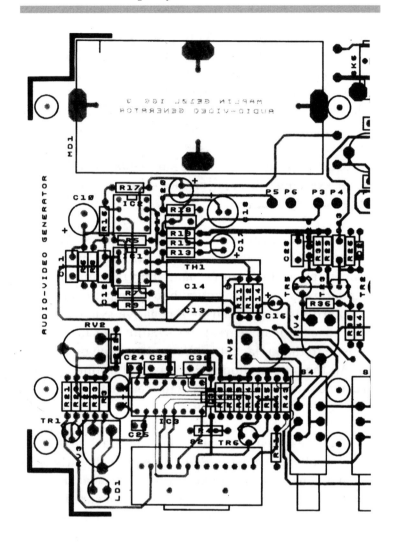

Figure 2.4   Layout of the PCB

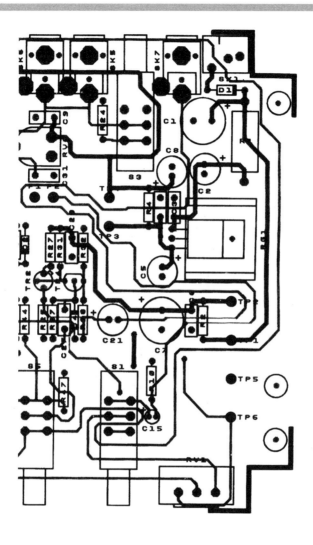

Figure 2.4 Continued

XT1 and the bead thermistor TH1 ensure that you don't over heat them, while making sure that they are firmly on the surface of the board. Next install the twelve pins at TP1 to TP6 and P1 to P6 ensuring that you push them fully into the board.

The RF modulator MD1 depends for its electrical screening and mechanical support on four large, flat solder tags, which insert through the four slots provided in the PCB. To secure the unit onto the PCB simply twist the fixing tags through 90 degrees. When soldering MD1 you must use an iron rated at 25 watts or more to ensure sufficient heating of the tag and solder pad on the PCB. The applied solder should then run freely round the joint until it fills the slot in the board. Make certain that all the wire connections of MD1 are soldered in their correct holes and not touching the metal screening.

Next install the push switches S1, 3, 4 and 5 making certain that they are pushed down firmly on to the surface of the PCB. *Do not* fit the push buttons onto the switches at this stage. Before mounting rotary switch S2 and level control RV1 cut both shafts to a length of 12 mm.

Prepare switch S2 in the following manner:

● remove the nut and shake-proof washer,

● position the stop ring for seven ways,

● refit the washer and nut.

Using the two screws and spacers supplied with S2, install the switch making certain that it is fixed firmly onto the PCB. When fitting RV1 ensure that the solder tags go fully into the PCB.

Next mount the PCB connectors SK1, 5, 6 and 7 ensuring that they are pushed firmly against the board. When mounting the red LED, LD1, it must be 7 mm above the board and bent over at 90°, as shown in Figure 2.5. The LED has a short lead and a flat edge on one side of its case to identify the cathode (K) connection.

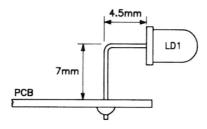

Figure 2.5   Fitting the LED

This completes assembly of the PCB. The remaining components are connected to the circuit board by cabling at a later stage. You should now check your work very carefully making sure that *all* the solder joints are sound. It is also very important that the solder side of the circuit board does not have any trimmed component leads standing proud by more than 3 mm, as this may result in a short circuit.

Final assembly

The PCB is designed to fit into an instrument case type 3502, with the front and back panels drilled as shown in Figure 2.6. When preparing the aluminium panels, the

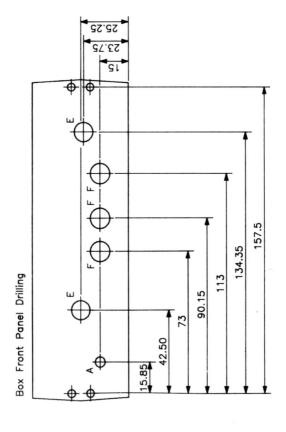

Box Front Panel Drilling

Figure 2.6   Front and back panel drilling

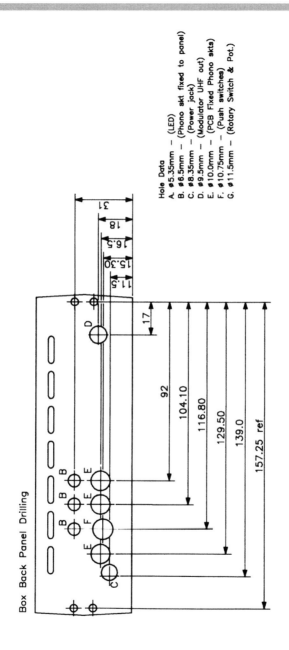

Box Back Panel Drilling

Hole Data
A. ⌀5.35mm — (LED)
B. ⌀6.5mm — (Phono skt fixed to panel)
C. ⌀8.35mm — (Power jack)
D. ⌀9.5mm — (Modulator UHF out)
E. ⌀10.0mm — (PCB Fixed Phono skts)
F. ⌀10.75mm — (Push switches)
G. ⌀11.5mm — (Rotary Switch & Pot.)

Figure 2.6 Continued

39

optional self-adhesive front and back stick-on panels can be used as a guide for checking the positioning of the holes. Having completed the drilling, at the same time clearing away any swarf, clean the aluminium panels and apply the trim by removing the protective backing. Carefully position and firmly push them down using a dry, clean cloth until they are securely in place.

Fit the front panel to the video pattern switch S2 and the audio level pot RV1 using the shake-proof washers and nuts provided with the controls. Secure the knobs so that their pointers are at the fully anti-clockwise position. Check that they travel smoothly round to the fully clockwise position, without scraping on the front panel. Next fit the four round push-buttons onto the plungers of S1, 3, 4 and 5.

Now you can carefully position the red LED through its hole in the front panel, then fit the rear panel using the side-chassis supports and screws supplied with the case. Lower the unit into the bottom half of the case ensuring that the panels slide smoothly into place and all the fixing holes in the PCB line up with the mounting points. Using six number $4 \times \frac{1}{4}$ in self-tapping screws secure the PCB to the case. Finally, mount the three phono sockets SK2, 3 and 4 on to the back panel.

Wiring

The total amount of wiring has been kept to a minimum by using PCB mounted switches and connectors leaving only three off-board components. Included in the kit is a half metre length of miniature screened audio cable which should be ample to make up the three links between P1 to P6 and SK2 to SK4, as shown in Figure 2.7.

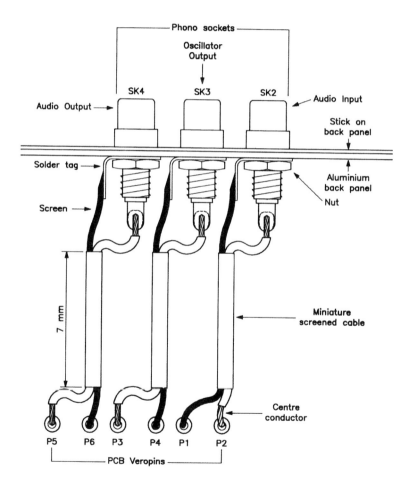

Figure 2.7 Wiring

D.C. testing

The d.c. tests can be made with a minimum of equipment. You will need a multimeter and a regulated +12 V d.c. power supply capable of providing at least 250 mA. The readings were taken from the prototype using a digital multimeter, some of the readings you obtain may vary slightly depending upon the type of meter employed.

Double check that none of the ICs has been fitted into the sockets on the board. The first test is to ensure that there are no short circuits before you connect the power supply. Set your multimeter to read *ohms* (Ω) on its resistance range, connect the test probes to TP5 and the anode of D1. With the probes either way round a reading greater than 500 Ω should be obtained.

Next monitor the supply current; set your meter to read d.c. mA and place it in series with the positive line of the power supply. With the power supply on a current reading of approximately 53 mA should be seen and the red LED, LD1, should light up. Turn off the supply and install the ICs making certain that all the pins insert into their sockets properly and the pin 1 marker on the IC package is at the notched end of the socket. Power up the unit and observe the current reading which should now be approximately 200 mA.

Remove the test meter from the positive supply and set it to read d.c. volts. All voltages are positive with respect to ground, so connect your negative test lead to the ground test point TP5. Before taking any readings set the PCB presets and the front panel controls to the following positions:

- RV1 (audio level) set fully counter clockwise,

- RV2 to RV6 should be set to their half way position,

- S1, S3 to S5 buttons out,

- S2 (video pattern) position one (vertical lines).

When the generator is powered up voltages present on the PCB assembly should approximately match the following readings:

$$TP1 = +11.0 \text{ V}$$
$$TP2 = +10.9 \text{ V}$$
$$TP3 = +5.0 \text{V}$$
$$TP4 = +4.9 \text{ V}$$
$$TP6 = +5.1 \text{ V}$$

This completes the d.c. testing of the generator, now disconnect the multimeter from the unit.

Audio and video adjustments

To obtain the best results from the generator an oscilloscope and frequency counter should be used. However, good results can be achieved by simply observing the picture on a domestic TV.

The first test is for the audio oscillator and requires no adjustment. Place a 1.5 kΩ load resistor across the oscillator output socket SK3 and attach your oscilloscope and frequency counter. The readings should approximately match the following:

- the oscilloscope should show a sine-wave at 1.13 V peak-to-peak,

- the frequency counter should display 1.00 kHz.

Transfer the oscilloscope and frequency counter to the audio output socket SK4. With the audio source switch S1 in its out (INT) position and a 600 Ω load resistor across SK4, take the following readings:

- set the audio level control RV1 fully anti-clockwise. The oscilloscope should show no signal until this control is advanced,

- at its maximum setting a 1 kHz sine-wave at 4.8 V peak-to-peak should be observed,

- inject an audio signal of up to 1.13 V peak-to-peak at SK2,

- with RV1 set fully clockwise and S1 pushed in (EXT) the oscilloscope should display the amplified signal.

Now set the front panel controls as follows:

- video pattern switch S2 to position three (cross-hatch),

- video TV switch S4 out (INT),

- video monitor switch S5 out (NOR),

- audio source switch S1 out (INT),

- audio level RV1 to half way position.

Next connect a phono to coax adaptor lead from the RF output of the modulator to the aerial input of a UHF television set. Using a spare channel selector, tune to approximately channel 36, where you should find the crosshatch pattern on the screen and a 1 kHz tone from the speaker. If there is excessive buzzing on the sound channel adjusting the sub-carrier fine tune, RV6, should minimise this noise.

Connect a 75 Ω load resistor or a video monitor to the video output socket SK7. Attach the scope probe to SK7 and make the following adjustments:

● make sure that the crosshatch pattern is selected,

● ensure that the video switches S4 and S5 are in the out (INT) position,

● adjust RV5 to show a 1 V peak-to-peak signal on the oscilloscope,

● adjust RV3 so that the vertical lines are the same width as the horizontal lines on the TV or monitor screen,

● select the grey-scale pattern and adjust RV2. for peak white on the left hand side of the screen, down to black level on the right,

● with the video input termination switch S3 pushed in, feed a 1 V peak-to-peak signal to SK5 or 6,

● push in the video TV switch S4 (EXT) and adjust RV4 to show a 1 V peak-to-peak signal on the oscilloscope.

This completes the adjustments to the generator, now remove any output load resistors and connecting leads from the unit. Using the screws provided fit the lid of the box and the four rubber feet. The generator is now ready for use.

Video and TV projects

Using the generator

To obtain optimum sound and picture quality you must use a regulated 12 volt power supply, capable of providing up to 250 mA. A suitable power supply would be mains adaptor YB23A.

The generator has two audio outputs. Socket SK3 provides a low distortion sine-wave at a fixed level, while SK4 produces an amplified signal with a variable level controlled by RV1. This output can be switched by S1 to an external audio source applied to SK2.

When the video TV switch S4 is set to internal (INT) the test patterns are selected and appear at the video output socket SK7. However, when this switch is pushed in, signals applied to the video input sockets SK4 and 5 will be selected. If the video monitor switch S5 is then pushed in (REV source) the test patterns will reappear at SK7, but not with the modulated TV output.

If the unit is placed in a video line already terminated by a 75 Ω load the termination switch S3 must be in the *off* position. The 1 MΩ impedance present at SK5 and 6 will contribute little to the loading of a 75 Ω line, so no attenuation of the signal should be detected.

The unit can be used as an audio/video modulator receiving its signals from a camera, VCR, or almost any composite video source. To achieve this the video TV switch S4 and the audio switch S1 must be pushed in (EXT). The audio level control RV1 should then be set for the correct sound level received by your TV.

Audio/video generator

Power supply voltage:	10 V to 15 V d.c.	
Supply current at 12 V:	195 mA	
Audio oscillator		
Waveform:	Sine	
Frequency:	1 kHz ±2%	
Output level:	400 mV r.m.s. 1k5 Ω load	
Distortion:	0.015% THD	
External audio		
Input level:	400 MV r.m.s.	
Input impedance:	5 kΩ	
Bandwidth ±3 dB:	6 Hz to 1.7 MHz	
Bandwidth ±6 dB:	4 Hz to 2.1 MHz	
Output level:	1.7 V r.m.s. 600 Ω load	
Distortion:	0.025% THD	
Video generator:	Black and white	
CCIR timing:	Line frequency	15.625 KHz (64 μs)
	Field frequency	50 Hz (20 ms)
Video patterns:	Vertical lines	16 visible
	Horizontal lines	18 visible
	Crosshatch	1.4:1 aspect ratio
	Dots	
	Grey-scale	8 steps
	Blank raster	
	White raster	
External video:	Colour or black and white	
Bandwidth:	7 MHz	
Input level:	1 V peak-to-peak	
Input impedance:	1 MΩ (no termination)	
	75 Ω (terminated)	
Output level:	1 V peak-to-peak 75 Ω load	
RF output:	Channel 36 (591.5 MHz)	
Sound sub-carrier:	6 MHz	
Output impedance:	75 Ω	

Table 2.1   Specification of prototype

Video and TV projects

Audio/video generator parts list

Resistors — All 1% 0.6 W metal film (unless specified)

R1	10 Ω 1 W wirewound	1	(C10R)
R2,4	10 Ω	2	(M10R)
R3,22	470 Ω	2	(M470R)
R5,6,29	4k7	3	(M4K7)
R7,43	270 Ω	2	(M270R)
R8,11,32, 41,47	820 Ω	5	(M820R)
R9,12,14	15 k	3	(M15K)
R10,18	10 k	2	(M10K)
R13,15, 21,30, 33,44	1 k	6	(M1K)
R16	8k2	1	(M8K2)
R17	27 k	1	(M27K)
R19,45	100 Ω	2	(M100R)
R20	150 Ω	1	(M150R)
R23,36	1k5	2	(M1K5)
R24	75 Ω	1	(M75R)
R25	330 Ω	1	(M330R)
R26,37	1 M	2	(M1M)
R27,35, 46	82 Ω	3	(M82R)
R28,31,48	3k3	3	(M3K3)
R34,38, 39,40	1k8	4	(M1K8)
R42	2k7	1	(M2K7)
RV1	10 k pot lin	1	(FW02C)
RV2	4k7 hor encl preset	1	(UH02C)

RV3	100 k hor encl preset	1	(UH06G)
RV4,5	1 k hor encl preset	2	(UH00A)
RV6	2k2 hor encl preset	1	(UH01B)
TH1	R53 bead thermistor	1	(FX62S)

Capacitors

C1,7	1000 µF 16 V PC elect	2	(FF17T)
C2,5,8, 18,21	220 µF 16 V PC elect	5	(FF13P)
C3,4,6,9, 11,12,19, 22,23, 27–30,31	100 nF disc	14	(YR75S)
C10	100 µF 16 V minelect	1	(RA55K)
C13,14	10 nF 1% polystyrene	2	(BX86T)
C15,16	4µ7F 35 V minelect	2	(YY33L)
C17,20	47 µF 16 V minelect	2	(YY37S)
C24	22 pF ceramic	1	(WX48C)
C25	10 pF ceramic	1	(WX44X)
C26	1800 pF ceramic	1	(WX71N)

Semiconductors

D1	1N4001	1	(QL73Q)
D2–4	1N4148	3	(QL80B)
LD1	LED red	1	(WL27E)
RG1	L7805CV	1	(QL31J)
TR1,6	BSX20	2	(QF32K)
TR2,5	BC179	2	(QB54J)
TR3,4	BF244A	2	(QF16S)
IC1	LF353	1	(WQ31J)
IC2	LF351	1	(WQ30H)
IC3	ZNA234E	1	(UK83E)

Video and TV projects

Miscellaneous

	constructors' guide	1	(XH79L)
	PC board	1	(GE10L)
XT1	2.5 MHz crystal	1	(UK82D)
SK1	2.5 mm d.c. pwr skt PCB	1	(FK06G)
SK2,3,4	phono chassis skt	3	(YW06G)
SK5,6,7	PCB phono skt	3	(HF99H)
S1,3,4,5	2-pole latchswitch	4	(FH67X)
S2	1 x 12 R/A rotary PCB	1	(FT56L)
MD1	UM1286 modulator	1	(BK66W)
	pins 2145	1 pkt	(FL24B)
	min screened	1 mtr	(XR15R)
	KB4 knob	2	(RW87U)
	latchbutton knob sm black	4	(BW13P)
	8-pin DIL socket	2	(BL17T)
	16-pin DIL socket	1	(BL19V)
	M10 pot nut	1 pkt	(FP06G)
	M3 x 10 mm isobolt	1 pkt	(HY30H)
	M3 isonut	1 pkt	(BF58N)
	M3 isoshake	1 pkt	(BF44X)
	vaned heatsink plas pwr	1	(FL58N)

Optional (not in kit)

case	1	(YN33L)
300 mA a.c. adaptor regulated	1	(YB23A)
stick on front panel	1	(JP03D)
stick on back panel	1	(JP02C)
no. 4 x $^1/_4$ in self tapping screw	1 pkt	(FE68Y)
phono/coax video lead plug	1	(FV90X)

All of the above items (except Optional) are available as a kit (LM98G)

3 Infra-red video link — part 1

The infra-red video link allows transmission of mono-chrome composite video over a distance of up to 100 metres. The system makes use of large Fresnel lenses to focus the infra-red energy at much greater distances than are possible using standard techniques.

The complete link is described over this and the next chapter. Part 1 of this pair of chapters gives description and constructional details of the transmitter. Part 2 (chapter 4) similarly covers the receiver.

Video and TV projects

Circuit description

Figure 3.1 shows the circuit diagram of the transmitter. The power supply is connected via SK1. Diode D1 provides reverse-polarity protection. The circuit requires two separate supply rails: a 12 V, high current supply; and a 5 V, low current supply. Regulator RG1, with its associated components, provides the regulated 5 V supply for IC1. A composite video signal, at a standard 1 V peak-to-peak level, is applied to the circuit between P1 and P2.

A video signal essentially consists of three parts; synchronisation (sync) pulses, a luminance (brightness) signal, and chrominance (colour) information. The circuit makes use of only the sync and luminance components; for practical purposes, the chrominance signal can be ignored. Separating the sync and luminance signals allows each of these to be handled separately, as each requires different processing.

Power supply voltage	12 V to 14 V
Power supply current (quiescent)	150 mA at 12 V
Range	Up to 100 m
PCB dimensions	83 x 83 mm approximately
Focal length of lens	270 mm approximately
Peak infra-red wavelength	940 nm
Composite video input	1 V peak-to-peak
Video bandwidth	4 MHz

Table 3.1   Specification of prototype

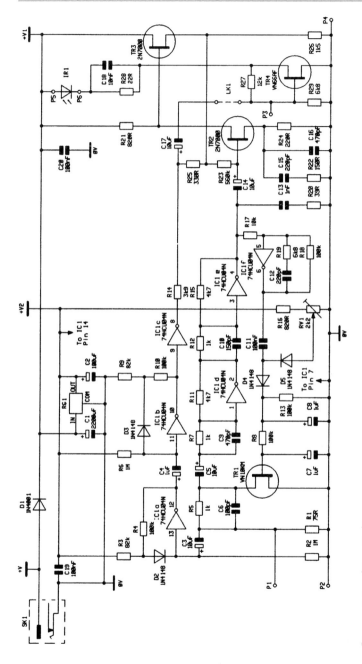

Figure 3.1 Circuit diagram

The high frequency luminance signal is separated and processed by IC1d and e. It is necessary for the overall gain of the circuit to peak at high frequencies to compensate for the poor response of the infra-red emitter at higher frequencies. Capacitors C6, C9 and C10 provide high frequency peaking.

The sync signal is processed by IC1a, b and c to remove the high frequency luminance signal, and to produce clean frame and line sync pulses.

IC1f forms part of an AGC (automatic gain control) amplifier. The output from this stage is rectified by D4 and fed to TR1, which maintains a relatively constant signal in the system, helping to prevent overloading. Preset resistor RV1 adjusts the AGC level, and indirectly controls the output level of the transmitter.

The sync and luminance signals are recombined after processing, and are fed to output transistor TR4, which drives the high power infra-red emitter, IR1. This is positioned at approximately the focal length of a large Fresnel lens such that the infra-red energy is focused to infinity.

Construction

After identifying the components and ensuring that all are present, insert and solder them onto the PCB, following the instructions given below and referring to the legend shown in Figure 3.2.

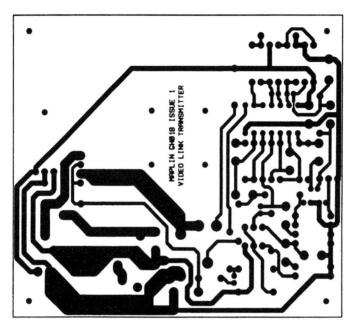

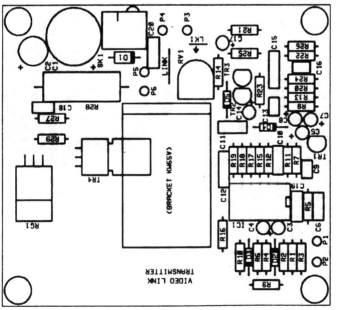

Figure 3.2 PCB legend and track

Video and TV projects

Start assembly by fitting the resistors, as these are fairly low profile components which may be difficult to fit at a later stage in construction. When fitting the IC socket, make sure that the notch at one end of the socket corresponds with that on the PCB legend; do not fit the IC until all other components have been soldered in place.

The PCB pins should be fitted next, these being inserted into the PCB and then pressed home using a hot soldering iron. When the pins are heated to the correct temperature, very little pressure is required to push them into place. There are two wire links on the PCB, and both of these are fitted in normal operation. Link LK1 connects the output of the driver stage to the power output stage and may be omitted if it is required to drive the power output stage directly via P3 and P4, as may be necessary in some non-standard applications. For standard operation using composite video, however, LK1 must be fitted.

Next, fit the capacitors, remembering to fit the electrolytic capacitors the correct way around. Note that capacitors C3 and C4 share the same + symbol on the board legend. All semiconductors must also be fitted with correct polarity, as shown on the PCB. The infrared transmitter diode is mounted on a metal bracket, which in turn is mounted on the component side of the PCB, as shown in Figure 3.3. The diode is mounted through the hole in the bracket; this arrangement should provide a relatively tight fit, and the small sachet of epoxy glue supplied with the kit can be used to hold the diode even more securely in position. Four nuts and bolts are used to fasten the bracket to the PCB. The diode is wired onto P5 (anode) and P6 (cathode) on the PCB, us-

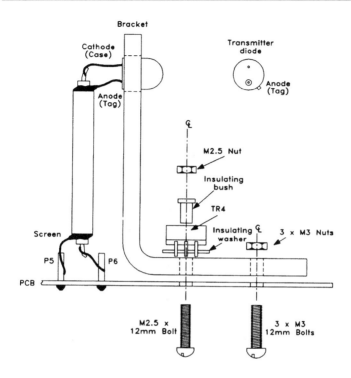

Figure 3.3 Mounting the emitter diode and diode bracket

ing the screened cable supplied. The leads of the diode should be cut to 10 mm and bent in such a way as to reduce the possibility of shorting. Sleeving may also be used if necessary, but the leads of the diode are comparatively rigid and should stay apart once correctly positioned.

The transistors (and regulator RG1) are fitted so that their cases correspond with the relevant outlines on the PCB. A hole is provided on the PCB, allowing the regulator to be bolted to the PCB using an M3 nut and bolt;

57

Video and TV projects

however, this is optional and it is not essential to do this. Transistor TR4 uses the diode bracket as a heatsink to aid heat dissipation, and is mounted as shown in Figure 3.4, using the M2.5 nut and bolt supplied in the kit.

Take note — Take note — Take note — Take note

The transistor needs to be insulated from the bracket with an insulating bush and a greaseless washer.

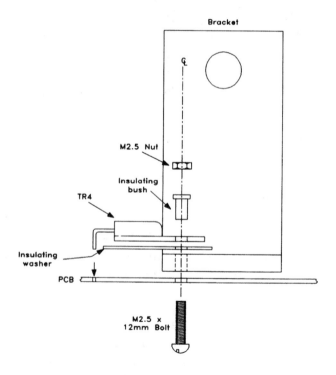

Figure 3.4   Mounting TR4 onto the bracket

Testing

Before applying power to the circuit, it is recommended that you double-check your work to ensure that all of the components are fitted correctly, and that there are no dry joints or solder bridges. Power supply connections are made via a PCB-mounted 2.5 mm power socket; the outer part of the power plug should be connected to 0 V, and the inner part to +V. The circuit requires a 12 V to 14 V d.c. power supply capable of supplying at least 500 mA. Although the module has its own on-board regulation and decoupling components, it is recommended that a power supply with a suitably smooth output is used to prevent any unwanted modulation of the supply rails. A suitable power supply for the unit is YZ21X. The composite video input signals are applied to P1 (input), and the return (ground) to P2 (0 V), via the BNC input socket. The input signal level should be approximately 1 V peak-to-peak.

It is not possible to test the unit fully without the appropriate infra-red receiver, or a full set of test equipment. The receiver is detailed in the next chapter. However, if you have an oscilloscope set up to display a video signal, it *is* possible to make basic tests on the unit. Apply a video signal to the input, and monitor the signal between P3 and ground (P4). Set preset RV1 to the centre of its travel, as marked by the arrow on the PCB legend. The oscilloscope should display a similar waveform to that of the original source video signal, although the levels will probably be different. The waveform displayed is that of the drive signal to the power output stage. If the signal on P6 is monitored, a somewhat compressed version of the signal should be displayed on the scope screen; this corresponds to the voltage across IR1. Although these tests are relatively simple, they allow basic operation of the circuit to be assessed. Setting of preset RV1 depends on the location of the transmitter and receiver, and should be left until both units are in place and the system is up and running.

Housing

An undrilled case is supplied in the kit, to house the transmitter. The drilling details are shown in Figure 3.5. The box has slots to hold the Fresnel lens in place. In each case, the PCB is mounted using 4 x M3 nuts and bolts, $^1/_4$ in spacers are used to position the PCB at the correct height in the case. The PCB is mounted such that the vertical part of the *diode bracket* is positioned 270 mm from the lens (as accurately as possible); this is approximately the focal length of the lens and has been found to provide the optimum range. Figures 3.6 and 3.7 show how to mount the lens and PCB in the case (see note below). The input (BNC) socket, SK2, is mounted on the rear panel, and is wired to the PCB. The power socket (SK1) does not require any additional wiring, being PCB mounted. It is, however, necessary to drill a suitable hole in the rear panel of the box to allow a 2.5 mm power plug to be inserted.

The Fresnel lens, as supplied, is too large to fit into the end of the box supplied in the kit, and a box to hold a lens of this size would not be practical in many cases. It is therefore necessary to cut the lens to the correct size to fit into the box, and this is 103 mm x 103 mm. It is

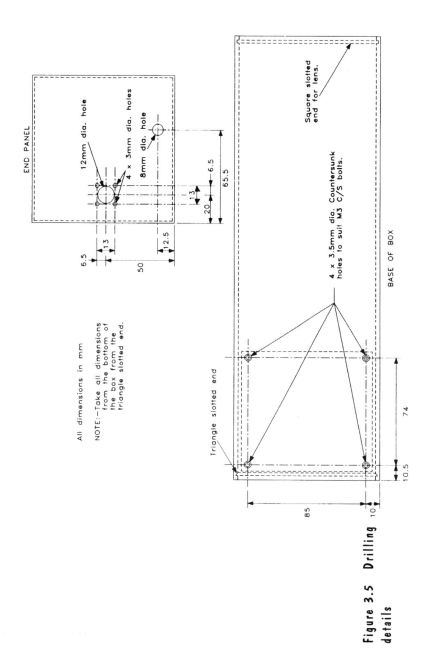

END PANEL

12mm dia. hole

4 × 3mm dia. holes

8mm dia. hole

6.5

6.5

13

65.5

13

20

50

12.5

All dimensions in mm

NOTE:—Take all dimensions from the bottom of the box from the triangle slotted end.

Square slotted end for lens.

4 × 3.5mm dia. Countersunk holes to suit M3 C/S bolts.

BASE OF BOX

Triangle slotted end

74

10.5

85

10

Figure 3.5 Drilling details

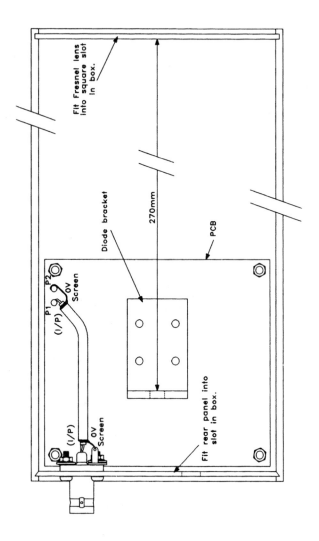

Fit Fresnel lens into square slot in box.

270mm

Diode bracket

PCB

P1 P2
(I/P)
OV
Screen

(I/P)
OV
Screen

Fit rear panel into slot in box.

Figure 3.6 Positioning the PCB and wiring SK2

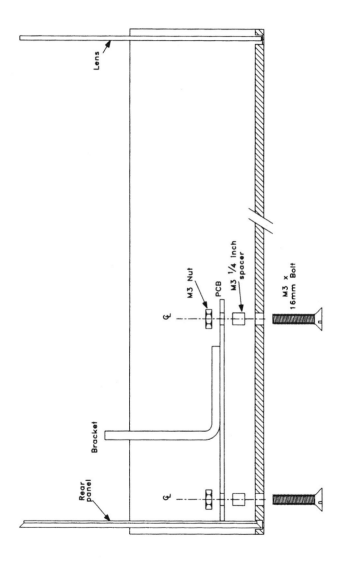

Figure 3.7 Mounting the PCB, lens and rear panel

important that the lens is cut as carefully and accurately as possible to ensure a correct fit. In addition, when trimming the lens down to size, material should be cut equally from the four sides so that the central point of the concentric rings is in the centre of the square, as shown in Figure 3.8. The box is not resistant to water and if the system is used outdoors, a degree of waterproofing is necessary. This may be achieved by smearing a thin layer of silicone rubber sealant around all of the box joints, including the grooves for the lens and the rear panel. A suitable sealant is YJ91Y. It is recommended that where possible, the unit is mounted in a sheltered location.

Mounting considerations

It is necessary to provide a secure mounting point for the transmitter and receiver units. This may be achieved in a variety of ways; one method, used for the prototype, makes use of small speaker stands (stock code GL18U), which are supplied in a pack of 2 with a selection of different types of bracket. The stands are particularly useful as they allow the bracket to be moved freely for alignment purposes, and also clamp securely in place when alignment is complete. There are obviously many different methods of mounting the transmitter and receiver units, and the most suitable method must be chosen for each individual situation.

It is important that the finished transmitter is *not* pointed directly at the sun, as the lens could focus the sun's rays onto the PCB, emitter diode or associated components,

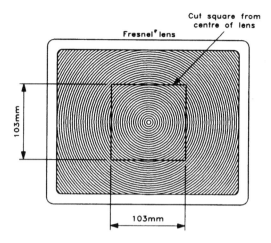

Figure 3.8 Cutting the Fresnel lens

generating heat and causing irreparable damage. This
consideration may dictate where the unit is mounted,
and in problem cases it may prove beneficial to devise a
protective hood for the front of the unit, to shield the
lens from direct sunlight.

Video and TV projects

Infra-red video link transmitter parts list

Resistors — All 0.6 W 1% metal film (unless specified)

R1	75 Ω	1	(M75R)
R2,6	1 MΩ	2	(M1M)
R3,9	82 kΩ	2	(M82K)
R4,8,10, 13,18	100 kΩ	5	(M100K)
R5,7,12	1 kΩ	3	(M1K)
R11,15	4k7	2	(M4K7)
R14	3k9	1	(M3K9)
R16,21	820 Ω	2	(M820R)
R17	10 kΩ	1	(M10K)
R19,29	6k8	2	(M6K8)
R20	39 Ω	1	(M39R)
R22	150 Ω	1	(M150R)
R23	560 kΩ	1	(M560K)
R24	220 Ω	1	(M220R)
R25	330 Ω	1	(M330R)
R26	1k5	1	(M1K5)
R27	12 kΩ	1	(M12K)
R28	22 Ω 3 watt wirewound	1	(W22R)
RV1	2k2 hor encl preset	1	(UH01B)

Capacitors

C1	2200 µF 16 V PC elect	1	(FF60Q)
C2	100 µF 16 V minelect	1	(RA55K)
C3,5, 14,17	10 µF 16 V minelect	4	(YY34M)

C4,7,8	1 µF 63 V minelect	3	(YY31J)
C6	100 pF ceramic	1	(WX56L)
C9,16	470 pF ceramic	2	(WX64U)
C10	150 pF ceramic	1	(WX58N)
C11,19, 20	100 nF disc	3	(YR75S)
C12,15	220 pF ceramic	2	(WX60Q)
C13	1000 pF ceramic	1	(WX68Y)
C18	10000 pF ceramic	1	(WX77J)

Semiconductors

IC1	SN74HCU04N	1	(UB04E)
RG1	L7805CV	1	(QL31J)
TR1	VN10KM	1	(QQ27E)
TR2,3	2N7000	2	(UF89W)
TR4	VN66AF	1	(WQ97F)
D1	1N4001	1	(QL73Q)
D2,3,4,5	1N4148	4	(QL80B)
IR1	I/R photo emitter	1	(KW66W)

Miscellaneous

P1–P6	pin 2145	1 pkt	(FL24B)
	14-pin DIL socket	1	(BL18U)
	PCB	1	(GH01B)
SK2	BNC square socket	1	(YW00A)
SK1	2.5 mm d.c. PCB pwr skt	1	(FK06G)
	I/R video case	1	(GL48C)
	Fresnel lens (large)	1	(KW60Q)
	bracket	1	(KW65V)
	TO220 insulator	1	(QY45Y)

Video and TV projects

TO66 plastic bush	1 pkt (JR78K)
low C cable	1 mtr (XR19V)
16 swg 1.6 mm TC wire	1 (BL11M)
M3 x 16 mm poziscrew	1 pkt (JC70M)
M3 steel nut	1 pkt (JD61R)
M2.5 x 12 mm steel screw	1 pkt (JY31J)
M2.5 steel nut	1 pkt (JD62S)
M3 x 12 mm steel screw	1 pkt (JY23A)
double bubble sachet	1 (FL45Y)
M3 x $^1/_4$ in spacer	1 pkt (FG33L)
instruction leaflet	1 (XK44X)
constructors' guide	1 (XH79L)

All of the above are available as a kit (LP59P)

4 Infra-red video link — part 2

The infra-red video link receiver circuit diagram is shown in Figure 4.1, while Figure 4.2 shows the PCB legend. The circuit can effectively be split into two parts; the pre-amplifier, and the main video processing circuit. The receiver uses two separate supply rails, in a similar way to that of the transmitter. The receiver lens focuses the infra-red energy onto photodiode PD1. High-frequency peaking is also required in the receiver, to compensate for the poor frequency response of the photodiode. IC1 amplifies the received signal from the photodiode, providing a gain that increases with frequency but falls off at around 4 MHz (the highest video frequency that can

Video and TV projects

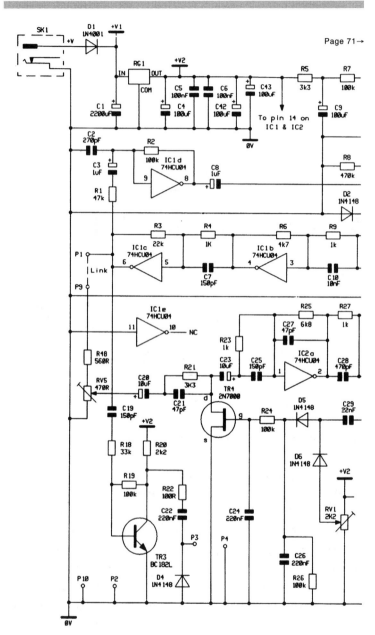

Page 71→

Figure 4.1   Circuit diagram

Infra-red video link 2

←Page 70

Page 72→

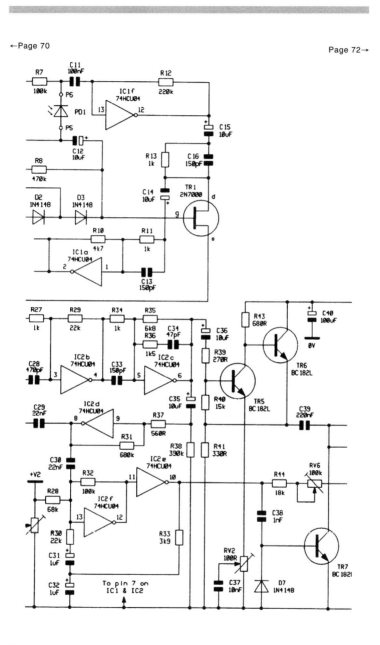

Figure 4.1   Continued

71

Video and TV projects

←Page 71

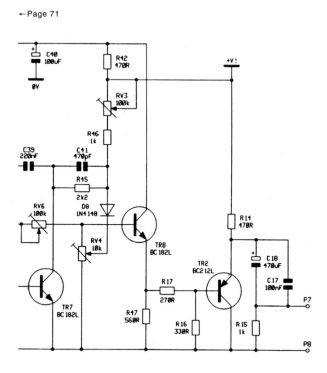

Figure 4.1   Continued

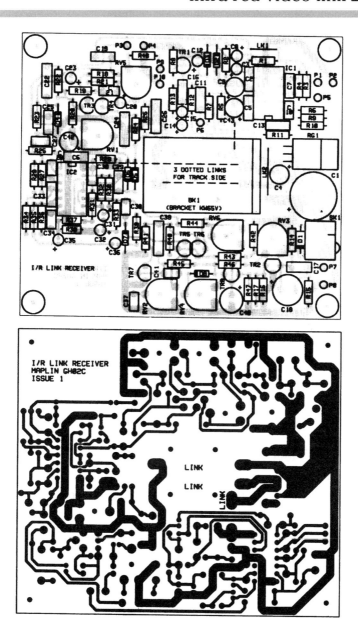

Figure 4.2   PCB legend and track

be resolved using this system). Part of the signal is fed back via IC1d and field effect transistor TR1, which serves as a simple automatic gain control (AGC) at the initial pre-amplification stage.

The signal, taken from IC1c, is then fed to the second stage of the circuit comprised of IC2 and associated components. Preset RV5 is configured as a level control, enabling the optimum signal level to be set for the next stage. IC2a, b and c act as an amplifier and filter stages; IC2d, TR4 and associated components forming a secondary AGC circuit. Preset resistor RV1 sets the level at which the AGC operates. IC2e and f act as a sync separator, which recovers the composite sync pulses and cleans them up. Preset resistor RV6 sets the final sync level. The luminance part of the signal is processed separately via TR5, TR6, TR7 and associated components, this stage largely rejecting the sync component of the signal. Preset RV2 limits the high frequency response of the signal, effectively providing a *sharpness* control. Both the sync and luminance signals are recombined at the base of transistor TR8. Preset resistor RV3 determines the black level, while RV4 sets the luminance level. The recombined signal is buffered by TR2, providing an output suitable to drive a standard video monitor.

An additional section of the circuit, made up from transistor TR3 and associated components, provides a rectified but unfiltered output derived from the pre-amplifier stage. This could be useful for reference purposes when the final transmitter and receiver units are physically aligned. This output is not calibrated, and is only intended to provide a reference.

Receiver construction

The majority of receiver construction techniques are very similar to those of the transmitter (described in the last chapter), and have therefore been omitted now.

The infra-red receiver diode (PD1) is mounted differently from the transmitter diode, as can be seen from Figure 4.3. The diode is held in place using epoxy resin adhesive, which is supplied in the kit. It is important that the leads of the diode do not touch the metal bracket. To this end, a small piece of green plastic filter is used to insulate the diode from the bracket. The leads of the diode should be cut to a length of approximately 10 mm and should be bent outwards to reduce the possibility of unwanted shorting. A short length of screened lead is used to connect the diode to P6 and 0 V.

Take note — Take note — Take note — Take note

The green filter material is used only as an insulator, and should not be used as a filter in front of PD1. The body of PD1 effectively sits on the upper edge of the filter, so that there is a clear line-of-sight path between it and the Fresnel lens.

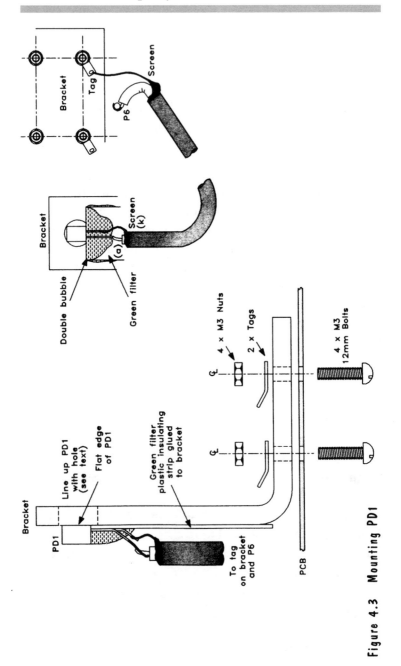

Figure 4.3 Mounting PD1

There are two wire links on the component side of the PCB. A component lead off-cut may be used for LK1; however, LK2 carries the main supply rail to regulator RG1, and must therefore be somewhat thicker. Suitable tinned copper wire is supplied in the kit for LK2.

In addition to the component side links, there are three links on the track-side of the PCB, each of which require special attention. To help with positioning these links (which are marked on the PCB) correctly, Figure 4.4 shows the links from the *track* side. It is important that these links are made up from insulated wire, to prevent them from shorting to other tracks, pads or component leads.

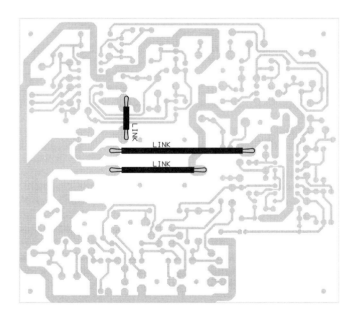

Figure 4.4   Track-side links

Video and TV projects

It is necessary to earth the diode bracket to prevent the introduction of noise into the pre-amplifier section of the circuit. The bracket is connected to ground via pin 5, using a short length of insulated wire. For optimum performance it is important that the length of this wire is kept short, and that wire of a suitably thick gauge (as supplied in the kit) is used.

The output from the pre-amplifier section of the circuit is available between P1 and P2, and is connected to the input of the video processing section (P9 and P10) via a length of screened cable. This approach is adopted to reduce the level of external noise introduced into the system, and it is recommended that the cable is kept as short as possible; this also applies to the other screened connections as well. The PCB wiring information is shown in Figure 4.5.

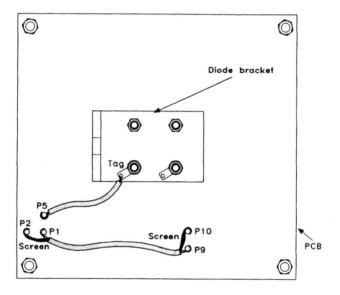

Figure 4.5   PCB wiring

Housing

An undrilled case is supplied with each of the transmitter and receiver kits. The box has slots to hold the Fresnel lens in place. In each case, the PCB is mounted using four M3 nuts and bolts, $1/_4$ in spacers being used to position the PCB at the correct height in the case. The PCB is mounted so that the diode bracket is positioned, as accurately as possible, at a distance of 270 mm from the lens. This is approximately the focal length of the lens, and has been found to provide optimum range (see Figure 4.6). Using the supplied case, a suitable spacing between the diode and lens will be achieved if the drilling details are followed closely (see Figure 4.7). Figure 4.8 illustrates how to mount the PCB in the case.

The Fresnel lens, as supplied, is too large to fit into the end of the box supplied in the kit, and a box to hold a lens of this size would not be practical in many cases. It is therefore necessary to cut the lens, so that it can be accommodated by the box, to dimensions of 103 x 103 mm (see Figure 4.9). Although the lens material is easy to cut using ordinary scissors, it is important to ensure that the lens is cut as carefully and as accurately as possible to ensure a correct fit.

The video output socket is mounted on the rear panel of the box, and is wired onto the appropriate pins on the PCB; the screened lead provided in the kit is used to wire it to the appropriate pins of the PCB.

The video lead should be screened to prevent unwanted radiation or external noise pick-up, which could create interference. The power socket is of the PCB mounting type, and does not require any external wiring.

Video and TV projects

Testing

Before the receiver is powered up, it is recommended that you check your work to make sure that there are no dry joints or solder bridges. In order to test the receiver fully, it is necessary to have the matching transmitter (available in kit form, stock code LP59P). There are 6 preset resistors on the PCB, which require setting up to suit each individual installation. Preset resistor RV1, which sets the AGC level, has a large effect on the final output level. The frequency response of the circuit is adjusted using RV2, which acts as a lowpass filter. Preset RV3 adjusts the black level, and RV4 sets the luminance level. Preset RV5 adjusts the input level to the video processing circuit, while RV6 sets the sync level. If you have access to an oscilloscope, you could monitor the output, adjusting the presets individually until a satisfactory output waveform is obtained; however, it is relatively easy to adjust the circuit by observing the picture and adjusting the appropriate presets until acceptable results are obtained. This *picture* method produces the same results (or even superior ones!), allowing the circuit to be adjusted to suit individual monitors. In most cases it is possible to align the unit without the box lid in place, in which case there will be no problem in gaining access to the presets; however, there are suitable access holes, shown as part of the drilling information, for the presets in the PCB.

For optimum picture quality, it is important that the optical alignment of the transmitter and receiver units is correct. Because the field of view is quite narrow,

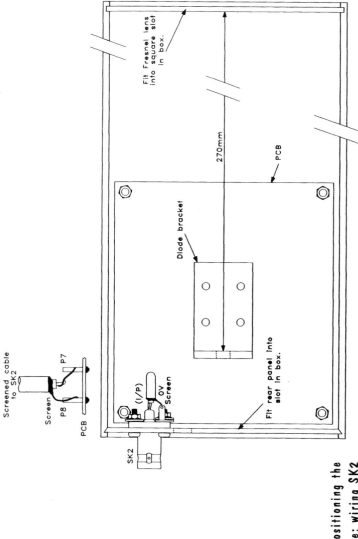

Fit Fresnel lens into square slot in box.

270mm

PCB

Diode bracket

Fit rear panel into slot in box.

Screened cable to SK2

P7

Screen

P8

PCB

(I/P)

0V Screen

SK2

Figure 4.6 Positioning the PCB in the case; wiring SK2

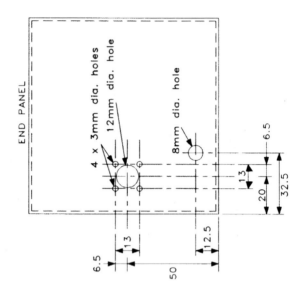

All dimensions in mm

NOTE:—Take all dimensions from the underside of the box from the triangle slotted end.

Figure 4.7 Drilling details

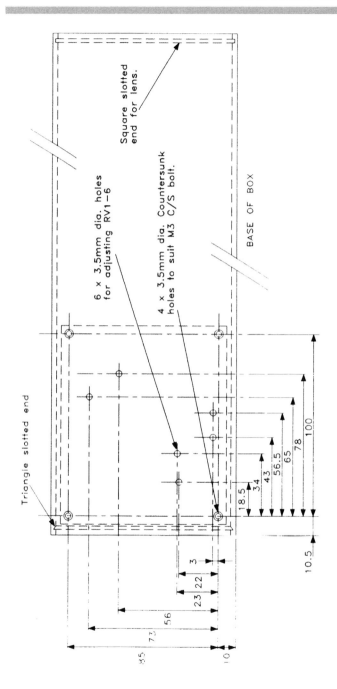

Square slotted
end for lens.

6 x 3.5mm dia. holes
for adjusting RV1-6

4 x 3.5mm dia. Countersunk
holes to suit M3 C/S bolt.

BASE OF BOX

Triangle slotted end

Figure 4.7 Continued

83

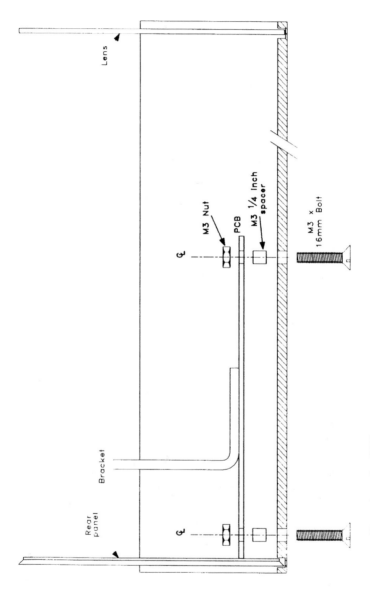

Figure 4.8   Mounting the PCB

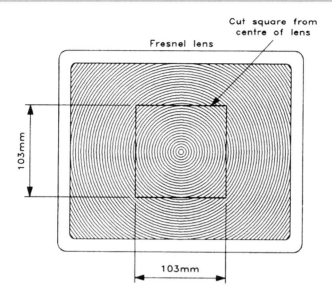

Figure 4.9 Cutting the Fresnel lens

alignment can be quite critical and some perseverance may be required when optically aligning the system. It is recommended that the system is initially aligned at short-range in order to make sure that both the transmitter and receiver are operating correctly. A d.c. voltage relating to the signal strength is available between P3 (output) and P4 (0 V); this output is designed to be used for optical alignment purposes only, and is not calibrated in any way. The output on P3 has a relatively high impedance, and so any measurements should be made with a multimeter (of sensitivity of at least 20,000 Ω/V).

Mounting considerations

As with the transmitter, it is necessary to provide a secure mounting point for the finished receiver unit. This may be achieved in a variety of ways; one method, used for the prototype, makes use of small speaker stands (stock code GL18U), which are supplied in pairs with a selection of different types of bracket. These stands are particularly useful as they allow the bracket to be moved freely for alignment purposes, and a clamp holds them securely in place when alignment is complete. There are obviously many different methods of mounting the transmitter and receiver units, and the most suitable method must be chosen to suit each individual situation.

If used outdoors, it is important that both the receiver and transmitter are suitably waterproof. Silicone rubber sealant (stock code YJ91Y) is useful for this purpose. It is essential to make sure that water does not enter the unit, as this may cause irreversible damage.

Applications

The system should provide reasonable performance if the guidelines mentioned are adhered to. The quality of picture is obviously not as good as that from a TV receiver, or a direct connection to a video monitor due to the reduced bandwidth. However, it is sufficient for general-purpose surveillance applications where fine detail is less important. It is particularly important that the transmitter and receiver are fixed securely in place, and are aligned in such a way that there is a clear line-of-sight path between the two units. It is essential to position the link so that the transmitter and receiver lenses are kept out of direct sunlight; the lenses can produce a very high temperature at the focal length in direct sunlight, and this may damage the unit. The area of the PCB around the focal length has been deliberately kept clear of high-profile components so that any damage is minimised. Nevertheless, direct sunlight focused into the unit can damage the receiver diode and surrounding components. In some cases, it may be found useful to fit a hood over the transmitter and receiver cases so that the lenses are shielded from sunlight. With regard to the infra-red link itself, it should be remembered that objects blocking the line of sight will prevent the system from operating.

The ambient environment (especially the light level) dictates the maximum range of the infra-red link. Under typical environmental conditions, a range of up to 100 m may be expected from the link; however, if the optical path is attenuated (for example, by fog or heavy rain), reduced range can be expected.

The system finds a wide variety of applications in the area of security, where it may be used to link a security camera to a video monitor. This is particularly useful between two buildings on the same premises. In this case, the infra-red link obviates the need for long external cable runs. Clearly, the system is only practical where the buildings are close enough together, and where there is a direct line-of-sight path between the transmitter and receiver. A similar arrangement inside a building will allow video information to be transmitted along corridors, for example.

Video and TV projects

Experienced constructors may wish to experiment with different modes of transmission. In addition to video transmission, it may also be possible (with the use of additional circuitry) to transmit other formats, such as digital information or speech. Possibly the simplest method of achieving this is to apply a carrier to the input of the transmitter, which is frequency modulated with the data to be transmitted. At the receiver end, the original data can be retrieved by demodulating the carrier. The use of frequency modulation (FM) reduces the possibility of signal degradation by interference from external sources. When using the link for purposes other than video transmission, the user may wish to bypass the video processing part of the circuit, as this may produce unwanted distortion. Access to the unprocessed signal is provided by P1 in the receiver. An input signal can be fed directly to the output stage of the transmitter via P3 (in this case link LK1 is not fitted on the transmitter PCB). Experimenters must ensure that any external apparatus connected to either the receiver or the transmitter in this way does not exceed the maximum ratings for any of the components in the circuit. In particular, signals applied to P3 of the transmitter should not be allowed to swing below 0 V, and should be limited to a maximum amplitude of +5 V.

If this system is used in an environment lit by mains lighting, you may find that its 50 Hz frequency modulates the beam. To reduce this problem, it is recommended that the receiver is pointed away from artificial light sources if possible. The effect of such modulation is usually only noticeable when the received signal is weak, and should not be a problem over short distances.

Power supply voltage	12 V to 14 V
Power supply current	200 mA
Video bandwidth	4 MHz approximately
Video output level	1 V peak-to-peak
System operating range	100 m maximum

Table 4.1   Specification of prototype

Infra-red video link receiver parts list

Resistors — All 0.6 W 1% metal film (unless specified)

R1	47 kΩ	1	(M47K)
R2,7,19, 24,26,32	100 kΩ	6	(M100K)
R3,29,30	22 kΩ	3	(M22K)
R4,9,11, 13,15,23, 27,34,46	1 kΩ	9	(M1K)
R5,21	3k3	2	(M3K3)
R6,10	4k7	2	(M4K7)
R8	470 kΩ	1	(M470K)
R12	220 kΩ	1	(M220K)
R14,42	470 Ω	2	(M470R)
R16,41	330 Ω	2	(M330R)
R17,39	270 Ω	2	(M270R)
R18	33 kΩ	1	(M33K)
R20,45	2k2	2	(M2K2)
R22	100 Ω	1	(M100R)
R25,35	6k8	2	(M6K8)
R28	68 kΩ	1	(M68K)
R31	680 kΩ	1	(M680K)
R33	3k9	1	(M3K9)
R36	1k5	1	(M1K5)
R37,47,48	560 Ω	3	(M560R)
R38	390 kΩ	1	(M390K)
R40	15 kΩ	1	(M15K)
R43	680 Ω	1	(M680R)
R44	18 kΩ	1	(M18K)

RV1	2k2 hor encl preset	1	(UH01B)
RV2	100 Ω hor encl preset	1	(UF97F)
RV3,6	100 kΩ hor encl preset	2	(UH06G)
RV4	10 kΩ hor encl preset	1	(UH03D)
RV5	470 Ω hor encl preset	1	(UF99H)

Capacitors

C1	2200 µF 16 V PC elect	1	(FF60Q)
C2	270 pF ceramic	1	(WX61R)
C3,8,31,32	1 µF 63 V minelect	4	(YY31J)
C4,9,42,43	100 µF 10 V minelect	4	(RK50E)
C5,6,11,17	100 nF 16 V minidisc	4	(YR75S)
C7,13,16, 19,25,33	150 pF ceramic	6	(WX58N)
C10,37	10 nF ceramic	2	(WX77J)
C12,14,15, 20,23, 35,36	10 µF 16 V minelect	7	(YY34M)
C18	470 µF 16 V PC elect	1	(FF15R)
C21,27,34	47 pF ceramic	3	(WX52G)
C22,24, 26,39	220 nF poly layer	4	(WW45Y)
C28,41	470 pF ceramic	2	(WX64U)
C29,30	22 nF ceramic	2	(WX78K)
C38	1 nF ceramic	1	(WX68Y)
C40	100 µF 16 V minelect	1	(RA55K)

Semiconductors

RG1	L7805CV	1	(QL31J)
IC1,2	74HCU04	2	(UB04E)
TR1,4	2N7000	2	(UF89W)

Video and TV projects

TR2	BC212L	1	(QB60Q)
TR3,5,6, 7,8	BC182L	5	(QB55K)
D1	1N4001	1	(QL73Q)
D2,3,4,5,6, 7,8	1N4148	7	(QL80B)
PD1	infra-red photodiode	1	(YH71N)

Miscellaneous

SK1	2.5 mm d.c. PCB pwr skt	1	(FK06G)
SK2	BNC square socket	1	(YW00A)
	Fresnel lens (large)	1	(KW60Q)
	I/R video case	1	(GL48C)
	bracket	1	(KW65V)
	PCB	1	(GH02C)
P1–P10	2145 pin	1 pkt	(FL24B)
	14-pin DIL socket	2	(BL18U)
	M3 x 16 mm poziscrew	1 pkt	(JC70M)
	M3 x 12 mm steel screw	1 pkt	(JY23A)
	M3 steel nut	1 pkt	(JD61R)
	M2.5 x 12 mm steel screw	1 pkt	(JY31J)
	M2.5 steel nut	1 pkt	(JD62S)
	M3 x 1/4 in spacer	1 pkt	(FG33L)
	miniature coax	1 mtr	(XR88V)
	epoxy resin sachet	1	(FL45Y)
	1.6 mm 16 swg TC wire	1	(BL11M)
	16/0.2 mm 10 m blk wire	1 pkt	(FA26D)
	M3 isotag	1 pkt	(LR64U)
	filter green	1	(FR33L)
	instruction leaflet	1	(XT45Y)
	constructors' guide	1	(XH79L)

All of the above are available as a kit (LP99H)

5 VHS recorder alarm

Described here is a portable, self-contained, alarm system disguised as a video cassette tape which detects movement from any pre-determined position. The cassette can be inserted into front or top loader VHS video players and will give an audible warning if the machine is moved or the cassette ejected. In addition, the cassette could be placed on top of a video, TV or hi-fi or mixed with other tapes in a library situation and indeed in any position where a video cassette will not appear obtrusively out of place.

The module is powered by 9 V PP3-sized batteries, and either dry-cell or ni-cad types may be used. If a ni-cad battery is fitted, the module has a constant current charger (approximately 10 mA) circuit included which requires a separate 12/15 V d.c. supply for recharging the battery in situ. With the addition of a case mounted socket wired to the charging circuits, the ni-cad can be recharged at any time without taking the cassette apart, as would be the case when using dry cells. Ordinary battery life expectancy should be quite long, and when *armed* the module's quiescent current is some 0.00002 A (20 µA) at 9 V d.c.

When the module is first switched *on* an LED lights for approximately 10–12 seconds, this being the *arm delay* time-out period. After this time, the LED extinguishes and the system is *armed* for detecting movement. When moved, the module delays the alarm sounders for approximately 6 seconds then triggers two electronic buzzers, which will sound continuously until either the module is switched off, or the battery supply runs down.

Operating principle

Movement detection using mercury tilt switches relies on a *make* or *break* operation which limits the device to a single plane of movement in one direction only. The device could only be used by correct positioning to begin with! For a portable alarm system, the criteria for movement detection has to be for 360° rotation in all directions no matter where the device is placed, or at what angle. To achieve this effect, two tilt switches are used, one operating vertically and the other operating horizontally. By sensing a change of state from either sensor, rather than looking for make or break action alone, the system may be placed at any angle at the outset and movement from this position into a different plane can then be detected. Such a system can be made quite precise by accurately positioning the sensors, or less sensitive just by altering the incidence angle to suit requirements.

How it works

The two tilt switch sensors are shown in Figure 5.1 as S1 and S2, and each triggers a dual monostable, IC1 and IC2. IC1 has two monostables contained within the package, one of which is configured for positive edge triggering and the other is configured for negative edge triggering. Both trigger inputs are commoned at pins 4 and 11, and held high by R5. Triggering on the negative edge occurs upon S1 closing, and on the positive edge on S1 opening. Timing components R1 and C1 (R2, C2) determine the width of the triggered pulses which are output to D2 or D6 from IC1 pins 7 and 9 (IC2 and S2 circuitry is identical to IC1).

The four diodes D2, 3, 4 and 6 serve as a simple logic OR gate to trigger the latch formed around IC3. When a switch is activated, negative going trigger pulses, output from IC1/IC2, forward bias one of the OR gate diodes from R11. IC3 input pins 5 and 6 drop towards 0 V, while pin 4 (connected to pin 9 of a second NAND gate input) goes high. Providing pin 8 on this gate is also high, the output from pin 10 goes low.

The feedback resistor R11 connected between pin 10 and the latch input gate will hold the LATCH in this new state until power is removed. A turn on delay is included for

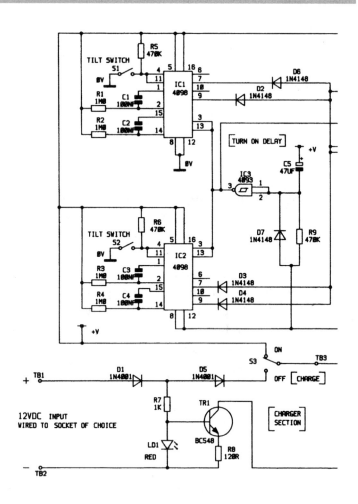

Figure 5.1   Video alarm circuit

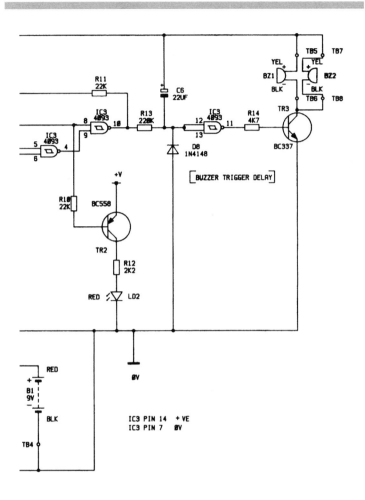

Figure 5.1   Continued

two reasons. Firstly, to prevent instant triggering when the power is switched on and secondly, to allow the user time enough to install the unit into a video player. At power on, capacitor C5 holds the inverter input pins 1 and 2 high while charging via R9. The inverter output pin 3 remains low at this time preventing IC1 and IC2 from being triggered and also disabling the latch. TR2 conducts and LED2 illuminates.

As C5 continues to charge up, current through R9 gradually decreases and hence the potential across R9 also decreases until the trigger threshold of the inverter is reached. The full CR time of a 47 µF capacitor and a 470 k resistor is some 22 seconds, but after 10 seconds, the inverter output snaps high thus enabling IC1/2, the latch circuitry, and turns off LED2. When triggered, IC3 pin 10 goes low and capacitor C6 discharges through R13, and the voltage at inverter input pins 12 and 13 slowly drops. This timing circuit allows a short, five second delay before the buzzers are activated, in other words while ejecting the unit from a video and switching off the power. The buzzers, BZ1 and BZ2 are used for loud output and modulating tone. Sound output level will inevitably be reduced when the unit is inside a video player, but is quite loud in open air.

If rechargeable PP3 batteries are to be used for the supply, then it is convenient to be able to recharge them in situ. This can only be performed with the *power-on* switch S3 in the *off* position, and the alarm becomes inoperative while charging. A 12 to 15 V d.c. supply connected to pin 1 and pin 2 operates LED1, and charging current flowing through the battery via S3 and TR1 collector is determined by R8. LED1 will not come on for a supply voltage of less than 12 V, and at 15 V will give maximum illumination. Some unregulated supplies may produce 15 to 18 V, but this is not a problem as R8 determines a constant current of 10 to 12 mA, suitable for ni-cad charging. D5 prevents the battery from discharging through R7 and TR1, with external supply removed, and D1 prevents reverse supply connections from damaging the generator and battery.

Construction

Refer to the constructors' guide supplied in the kit for component recognition and assembly techniques. Space on the PCB (Figure 5.2) is very tight and components should be mounted neatly and flat onto the board. LED1 and 2 should be pushed home as far as possible and not left standing up above the board. Take note of polarity markings on electrolytic capacitors and diodes, and do not fit sensors S1 or S2 at this stage. As the PCB is double-sided, all holes are therefore plated through, so before soldering components, recheck your work as errors are very difficult to correct on this type of board. IC holders must be fitted also!

Carefully solder all component leads and cut off the excess wire ends close to the board, *without damaging tracks while doing so.* Clean excess flux and solder splashes with a suitable PCB cleaner and inspect the module.

Refer to Figure 5.3 for sensor mounting. Insert the sensor leads and leave a clearance between sensor base and PCB of 5 mm. Solder both leads and gently bend the sen-

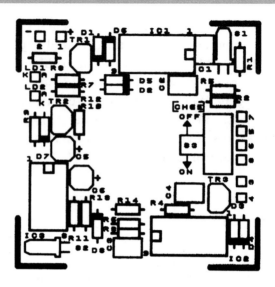

Figure 5.2   PCB layout

sor back over the legend ensuring that the metal case *does not short out on any lead*. As contact within the sensor is made by liquid mercury bridging internal contacts, S1 and S2 should be angled for more than (or less than) 20° to ensure fully open, or closed, contacts — whichever is required (see Figures 5.4 and 5.5).

Mounting the sensors horizontally flat onto the board will make the switches over sensitive to the slightest vibration, which, apart from being as desirable as it may seem, may actually lead to all sorts of false triggering problems. If the sensor is left standing vertically as in Figure 5.3, then it can be appreciated that a far greater degree of rotation is required to break contact. Therefore, adjust the sensor between 0 and 90° for best results. Manufacturer's data recommends 20° movement for activation.

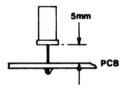

Figure 5.3   Mounting tilt switches

Position A

Position B

* See text.

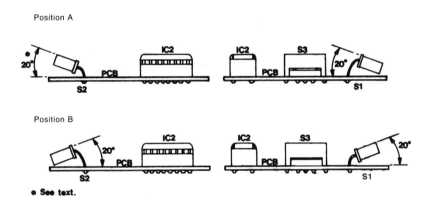

Figure 5.4   Adjusting tilt switches

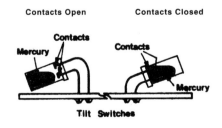

Figure 5.5   Tilt switch operation

Take the PP3 battery clip and cut the wires to a length of 6 cm. Strip and tin 2 mm from the ends and insert the black wire (negative) into PCB pin 4 and red wire (positive) into pin 3, as shown in Figure 5.6.

Do not reduce the length of the wires for each buzzer. Fit the yellow wire of one buzzer (BZ1) into pin 5, and the black wire into pin 6 of the PCB. For the second buzzer (BZ2), fit the yellow wire to pin 7 and the black wire to pin 8. Place the power switch, S3, into the *off* (charge) position.

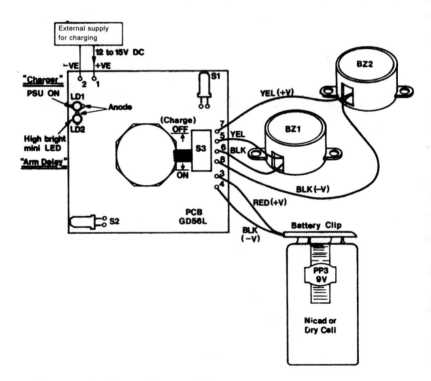

Figure 5.6   Connections to PCB

Testing

Lay the completed module, component side up, onto a non-conducting surface with both buzzers standing upright. Connect a PP3 battery to the clip and switch on by placing S3 in the *on* position. LED2 should glow brightly. If it is dim then LED2 and LED1 may be reversed. LED2 is a high brightness device and should not be replaced with standard mini LEDs. After approximately 10 seconds, LED2 will extinguish; the module is now armed. Do not move the module for a minute or so and ensure that the buzzers are not activated. Now move the module and after approximately 6 seconds the buzzers will sound continuously. LED2 does not operate during the trigger delay period, only during the turn on delay period. If all is well, adjust the sensors S1 and S2 for required movement detection and install the module into the blank case.

Case assembly

Refer to Figure 5.7 and Figure 5.8. Place the blank video cassette case bottom up and remove the five screws. Carefully separate both halves of the case, taking care not to lose the return spring fitted onto the front flap. Place the top section (large D shaped cut-outs) to one side, and position the base section as in Figure 5.8.

Mount four double-sided sticky pads around the spindle hole (left-hand side), remove the four paper strips from the pads and carefully place the PCB into position. The

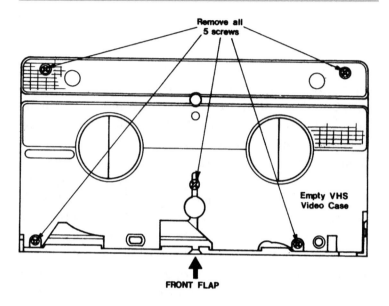

Figure 5.7 Opening the video case

hole in the PCB must be placed centrally over the hole in the case, otherwise the video drive spindle will foul on the board and may jam inside the machine! Final position of the board should be approximately 15 mm from the inside edge of the case. Figure 5.8 also shows the PCB hole centred on the case hole with 5 mm between edges. Press the PCB down onto the four pads to ensure good grip.

Position the two buzzers as shown ensuring the connecting wires do not cross over the spindle hole (for the reasons just mentioned), and use sticky pads to hold them in place. The battery should be placed in a position where it does not overlap the hole and this too is held with a sticky pad. If you require recharge facilities, then fit a socket suitable to take your supply plug into the case at this stage. Many battery eliminator supplies have multiple plug connectors from 2.1, 2.5 to 3.5 mm sockets. Use the smallest possible socket and fit it into the position shown above PCB pins 1 and 2 on Figure 5.8. *Do not* allow too much of the socket to protrude through the case as the majority of video machines usually have a vertical guide in this area! Suitable sockets can be found in the *connectors* section of the Maplin catalogue, if required. Reassemble the case halves and replace the five screws.

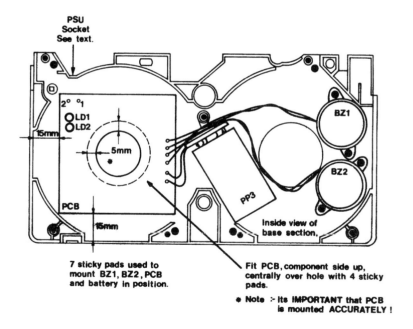

Figure 5.8   Video alarm assembly

Video and TV projects

Using the alarm

Powering up the alarm is achieved by inserting a finger through the base hole and the PCB. The unit is not automatically switched on when inserted into a video. During turn on delay, insert the cassette into a video player. If the cassette is immediately rejected, check the recording tabs are intact on the back edge of the box and replace with tape if they have been broken. If the alarm continuously triggers after timing out, then sensors S1 and S2 may need further adjustment. Any attempt to move the video player or eject the cassette, once it has been armed, will trigger the sounders after a few seconds. Alternatively, the cassette can be placed in a sleeve and left on top of a cassette library or a TV or wherever it is not likely to look out of place, giving other security uses.

VHS video alarm parts list

Resistors — All 0.6 W 1% metal film

R1,2,3,4	1 M	4	(M1M)
R5,6,9	470 k	3	(M470K)
R7	1 k	1	(M1K)
R8	120 Ω	1	(M120R)
R10,11	22 k	2	(M22K)
R12	2k2	1	(M2K2)
R13	220 k	1	(M220K)
R14	4k7	1	(M4K7)

Capacitors

C1,2,3,4	100 nF minidisc	4	(YR75S)
C5	47 µF 25 V PC electrolytic	1	(FF08J)
C6	22 µF 16 V PC electrolytic	1	(FF06G)

Semiconductors

IC1,2	4098BE	2	(QX29G)
IC3	4093BE	1	(QW53H)
LD1	red LED mini	1	(WL32K)
LD2	red LED hi bright	1	(WL83E)
TR1	BC548	1	(QB73Q)
TR2	BC558	1	(QQ17T)
TR3	BC337	1	(QB68Y)
D1,5	1N4001	2	(QL73Q)
D2,3,4, 6,7,8	1N4148	6	(QL80B)

Video and TV projects

Miscellaneous

BZI,2	buzzer	2	(CR34M)
S1,2	tilt switch	2	(FE11M)
S3	R/A SPDT slide	1	(FV01B)
	PC board	1	(GD56L)
	veropin 2145	1 pkt	(FL24B)
	PP3 battery clip	1	(HF28F)
	quickstick pads	1 strip	(HB22Y)
	14-pin DIL skt	1	(BL18U)
	16-pin DIL skt	2	(BL19V)
	VHS video case (blank)	1	(YP27E)
	constructors' guide	1	(XH79L)
	instruction leaflet	1	(XK60Q)

Optional (not in kit)

PP3 battery	1	(FK62S)
PP3 ni-cad battery	1	(HW31J)
power socket	if req	see text
300 mA d.c. PSU	1	(XX09K)

All of the above (excluding Option items) are available as a kit (LM27E)

6 Low noise video pre-amplifier

The SL561C is a monolithic integrated circuit which can be used in a number of different low noise pre-amplifier roles. It contains nine very high performance transistors; and associated biasing components, see Figure 6.1. All this circuitry is held in an 8-pin DIL (dual-in-line) package as shown in Figure 6.2. The absolute maximum ratings and electrical characteristics of the IC are detailed in Table 6.1. Its high gain, low noise design makes it suitable for use in audio and video systems at frequencies up to 6 MHz. Noise performance is optimised for source impedances between 20 Ω and 1 k making the device suitable for use with a number of transducers including photo-conductive infra-red (IR) detectors, magnetic tape heads and dynamic microphones.

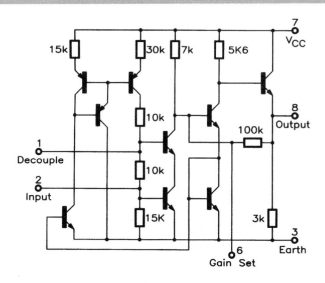

Figure 6.1   SL561C internal circuit diagram

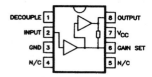

Figure 6.2   SL561C pin connections

The SL561C has an impressive specification, see Table 6.1, and its possible applications are too numerous for just one chapter to cover. However, it is hoped that the video pre-amplifier application here will provide a good starting point for your further experimentation.

Low noise video pre-amplifier

Characteristic	Min	Typ	Max	Units	Conditions
Voltage gain circuit	57	60	63	dB	pin 6 open
Equivalent input noise voltage	0.8			$nV / \sqrt{Hz}$	100 Hz to 6 MHz
Input resistance		3		kΩ	
Input capacitance		15		pF	
Output impedance		50		Ω	
Output voltage	2	3		V p–p	
Bandwidth		6		MHz	
Supply voltage		5	10*	V	
Supply current		2	3	mA	
Storage temperature	–55*		+125*	°C	
Operating temperature range	–55*		+100*	°C	

Test conditions: V_{cc} = 5 V; source impedance = 50 Ω; load impedance = 10 kΩ; T_{amb} = 25°C.

* Absolute maximum ratings.

Table 6.1 Absolute maximum ratings and electrical characteristics

Circuit description

A circuit diagram for the video pre-amplifier is shown in Figure 6.3. The power is applied to the PCB as follows, positive to the +5 V pin and negative to one of the 0 V ground pins. This supply must be within the range of +4.5 to +7 V and have the correct polarity otherwise damage may occur to the components. The positive d.c. supply is applied to pin 7 of IC1 (SL561C) and via R7 to the emitter of TR1 (BC328).

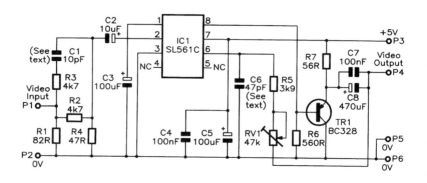

Figure 6.3 SL561C Video pre-amplifier circuit diagram

Main supply rail decoupling is provided by C5 (100 μF) with additional high-frequency decoupling provided by a 100 nF disc ceramic capacitor C4.

The video input to IC1 (pin 2) is made via an R/C attenuator network, R1 to R4 and C1, C2. Two criteria must be met if the external video device and the SL561C chip are to perform at their optimum level, namely matched input impedance and level. The incoming video signal is applied to the video input pin on the PCB and its ground is connected to its 0 V pin. The input impedance of the attenuator is significantly higher than the 75 Ω termination resistance required by most video equipment. To correct this a termination resistor R1 (82 Ω) is placed across the video input. If this is omitted the input impedance will increase to approximately 4k7 which will be useful if the video equipment is already terminated by some other device, i.e. a VCR or monitor. The video signal is then attenuated by resistors R2 and R4 to drive pin 2 of IC1 at the correct level, with capaci-

tor C2 providing the a.c. coupling. It is of great impor-
tance that this signal is at the correct level as it will
directly influence the harmonic content of the amplified
video signal, see Figure 6.4. To optimise the noise performance of the SL561C, the source impedance must be
between 20 Ω and 1 k see Figure 6.5. This parameter is
satisfied by R4 being a 47 Ω resistor. The video
attenuator stage is bypassed at high frequencies by a
low value (10 pF) ceramic capacitor, C1 and a 4k7 resistor, R3. This has the effect of boosting the upper
frequencies by a small amount, producing a slightly
sharper picture. If this enhanced image is not required,
then C1 and R3 may be omitted.

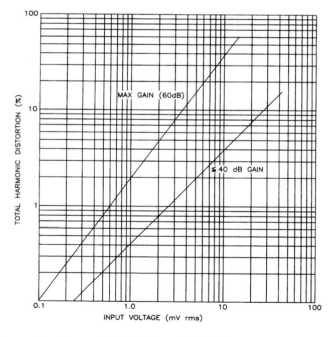

Figure 6.4   Harmonic distortion at 20 kHz

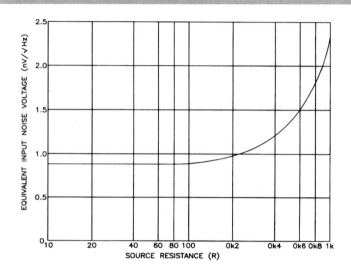

Figure 6.5 **Noise versus source impedance**

The low-frequency response is determined by the capacitors C2 and C3. C3 decouples an internal feedback loop on pin 1 and if its value is close to that of C2 an increase in gain at low frequencies occurs. For a flat response either make C3 less than 0.05 of C2 or make C3 five times, or more, greater than C2.

The bandwidth or upper cut-off frequency of the pre-amplifier can be reduced to any desired value by the capacitor C6 from pin 6 to ground. No degradation in noise or output swing occurs when this capacitor is used. A frequency response graph is shown in Figure 6.6 which depicts the differing −3 dB response curves produced by capacitors C1 and C6. Table 6.2 lists the components fitted/not fitted for the corresponding response curves shown in Figure 6.6.

Low noise video pre-amplifier

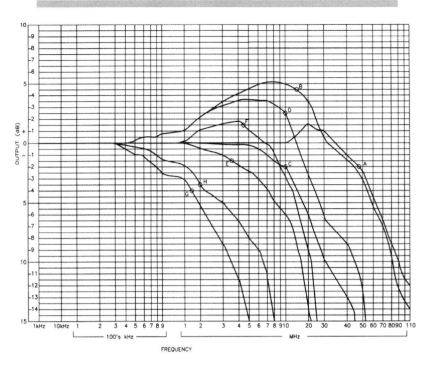

Figure 6.6 Typical frequency response of pre-amplifier

Power supply voltage:	5 V d.c.
Supply current:	40 mA
Bandwidth:	24 MHz (−3 dB)
Input level:	1 V p–p
Input impedance:	4k7 (no termination)
	75 Ω (terminated)
Output level:	2 V p–p
Output load impedance:	75 Ω

Table 6.2 Specification of video pre-amplifier prototype

Video and TV projects

The quiescent current of the output emitter on pin 8 of the SL561C (see Figure 6.1) is approximately 0.5 mA and only capable of driving moderately low impedance loads. As larger voltage swings are required into low impedance loads this current is increased by a 560 Ω resistor, R6, from pin 8 to ground. However, even this is not enough to drive the low impedance loads used by video systems. To accommodate this, a PNP transistor TR1 is used as a video output buffer. The video signal is a.c. coupled via C7 and C8 to the video output pin on the circuit board.

Provision is made to adjust the gain of the pre-amplifier by means of the combined values of R5 and RV1 between pin 6 and the output. Gain levels from as low as 10 dB and as high as 60 dB can be selected by altering this feedback resistor, see Figure 6.7. As the feedback increases (gain is reduced) around the output stage, instability problems can result if the bandwidth of the pre-amplifier is not reduced. Figure 6.7 shows the recommended values of C6 for each gain range. Since the input

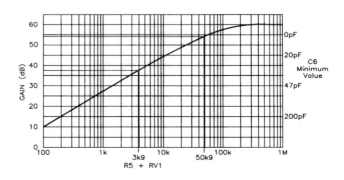

Figure 6.7 Gain set by R5 and RV1

stage is common emitter (without resistor or bypass capacitor), at values of gain less than 40 dB this input stage, rather than the output stage, determines the maximum output voltage swing. To keep distortion below 10%, the input signal level on pin 2 should be limited to 5 mV or less, see Figure 6.4.

PCB assembly

Removing a misplaced component can be difficult so double-check the type, value and polarity before soldering! The PCB has a printed legend that will assist you when positioning each item, shown in Figure 6.8.

The sequence in which the components are placed is not critical. However, the following instruction will be of use in making the task as straightforward as possible. It is

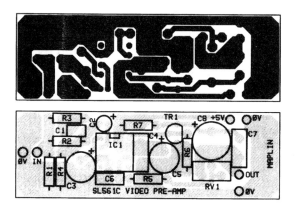

Figure 6.8 PCB legend and track

easier to start with the smaller components such as resistors (R1 to R7) followed by ceramic capacitors (C1, C4, C6, C7) and electrolytic capacitors (C2, C3, C5, C8). Polarity of the electrolytic capacitors is shown by a plus sign (+) on the PCB legend. However, the majority of electrolytic capacitors have the polarity designated by a negative symbol (−), in which case the lead nearest this symbol goes away from the positive sign on the legend.

Next install the preset resistor RV1 and set it to the fully anticlockwise position.

When fitting the 8-pin IC socket and the SL561C chip, ensure that you install them in the correct position, matching the notch with the block on the legend.

Finally, mount the BC328 transistor TR1, making sure that its outline aligns with the package outline on the legend. Install the six terminal pins ensuring that you push them fully into the board.

This completes assembly of the PCB. You should now check your work very carefully making sure that all the solder joints are sound. It is also very important that the solder side of the circuit board does not have any trimmed component leads standing proud by more than 3 mm, as this may result in a short circuit.

D.C. tests

The initial d.c. testing procedure can be undertaken using the minimum amount of test equipment. You will need a multimeter and a power supply capable of providing +5 V d.c. up to 100 mA. All the following readings are taken from the prototype using a digital multimeter; some of the readings you obtain may vary slightly depending upon the type of meter used!

Without any wires connected to the PCB terminal pins, carefully lay the PCB assembly on a non-conductive surface, such as a piece of dry paper or plastic. The first test is to ensure that there are no short circuits before connecting the d.c. power supply. Set your multimeter to read *ohms* on its resistance range and connect its two test probes to the +5 V and 0 V PCB pins. With the probes either way round, a reading greater than 10 kΩ should be obtained. If a significantly lower reading is registered, check solder joints and component leads, to make sure they are not shorting between tracks.

In the following test it will be assumed that the power supply used is set to provide a regulated +5 V d.c. supply. Select a suitable range on your meter that will accommodate a 100 mA d.c. current reading. Then place it in series with the positive supply to the PCB pin, and connect the negative supply to the 0 V pin, see Figure 6.9.

Turn on the power supply and observe the current reading, which should be approximately 40 mA. Switch off the power, then remove the test meter and connect the positive power line to the +5 V PCB pin.

Video signal wiring

All video signal wiring to and from the pre-amplifier module must be made using good quality screened 75 Ω

Video and TV projects

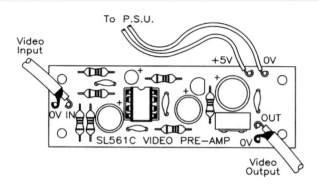

Figure 6.9   Wiring

low-loss coaxial cable (e.g., Maplin stock code XS32K), see Figure 6.9. The type of video connectors you use is a matter of personal choice and technical requirements.

Using the video pre-amplifier

To obtain the best results from the video pre-amplifier an oscilloscope should be used to set it up. However, good results can be achieved by simply observing the picture on a video monitor or a domestic TV with an AV input.

Useful applications are video pre-amplifiers, video buffers, and as part of a video distribution system. This could be a single video signal such as that from a satellite decoder, fed into the inputs of a number of SL561C boards, with the outputs to individual monitors.

120

A = 53 MHz (+1.5 dB peak at 20 MHz)	No C1 or C6
B = 50 MHz (+5 dB peak at 10 MHz)	C1 and no C6
C = 12 MHz	No C1 and C6 (47 pF)
D = 24 MHz (+3.5 dB peak at 7 MHz)	C1 and C6 (47 pF)
E = 6 MHz	No C1 and C6 (100 pF)
F = 10 MHz (+1.75 dB peak at 4 MHz)	C1 and C6 (100 pF)
G = 1 MHz	No C1 and C6 (470 pF)
H = 2 MHz	C1 and C6 (470 pF)

Table 6.3 Component options

SL561C Video pre-amplifier parts list

Resistors — All 0.6 W 1% metal film (unless specified)

R1	82 Ω	1	(M82R)
R2,3	4k7	2	(M4K7)
R4	47 Ω	1	(M47R)
R5	3k9	1	(M3K9)
R6	560 Ω	1	(M560R)
R7	56 Ω	1	(M56R)
RV1	47 k sub min enclosed vertical preset	1	(UH18U)

Capacitors

C1	10 pF metallised ceramic	1	(WX44X)
C2	10 μF 16 V sub min radial electrolytic	1	(YY34M)
C3,5	100 μF 16 V sub min radial electrolytic	2	(RA55K)
C4,7	100 nF 16 V mini disc ceramic	2	(YR75S)
C6	47 pF metallised ceramic (see text)	1	(WX52G)
C6	100 pF metallised ceramic (see text)	1	(WX56L)
C6	470 pF metallised ceramic (see text)	1	(WX64U)
C8	470 μF 16 V radial electrolytic	1	(FF15R)

Low noise video pre-amplifier

Semiconductors

| TR1 | BC328 | 1 (QB67X) |
| IC1 | SL561CDP | 1 (DB47B) |

Miscellaneous

8-pin DIL socket	1 (BL17T)
1 mm (0.04 in) single-ended	
PCB pin	1 (FL24B)
PCB	1 (GH61R)
instruction leaflet	1 (XU58N)
constructors' guide	1 (XH76L)

Optional (not in kit)

phono chassis socket	as req (YW06G)
phono screen plug	as req (HH01B)
UHF SO259 chassis	
socket	as req (BW84F)
UHF PL259 plug	as req (BW81C)
UHF PL259 reducer	
small	as req (BW82D)
UHF PL259 reducer	
large	as req (BW83E)
75 Ω BNC chassis socket	as req (FE31J)
75 Ω BNC plug	as req (FE99H)
PCB straight Scart socket	
(Peritel)	as req (JW34M)

Video and TV projects

PCB right Scart angle socket	as req (FV89W)
Scart plug	as req (FJ41U)

7 TV bar generator

If you are involved in sending television pictures up and down cables, or to and from VTRs and monitors, it is really necessary to be able to check if the pictures at the far end are being displayed correctly. Whilst the black and white test stripe that some VTRs give out in their test *mode* is useful, it is not normally possible to properly adjust a TV or monitor to *exactly* the right settings to display correct colour pictures.

Similarly, when servicing TVs, monitors, VTRs or other video equipment it is necessary to have a constant and stable colour test signal to be able to accurately set up the various preset controls. Over the last decade, the previously familiar colour test card has vanished from our TV screens much to the chagrin of TV and video engineers the length and breadth of the country. The younger readers of this magazine may not remember the days before the *Breakfast TV* and *24 hr TV* revolutions, prior to these it was easy to find at least one channel displaying a *test card*. Originally, the test card was literally placed in front of a camera, the most famous being the BBC's *Test Card F* which showed a picture of a young girl playing *noughts and crosses* on a blackboard which was in a circle in the centre of the screen with various squares and lines on the outside. Across the top was a bar showing the seven primary colours. Modern test cards are entirely electronically generated and usually personalised so that the signal can be identified.

Test signals on tap

For the engineer or technician an *on tap* colour test signal is needed, units to generate such a signal vary in complexity from an expensive broadcast standard electronic test card generator costing several thousand pounds to a much simpler and cheaper colour bar generator. Broadcast standard equipment is designed to the highest standards possible — after all the test equipment has to be better than the system under test.

A simple colour bar generator is more than adequate for most general setting up procedures, and when aligning video equipment a full test card provides *too much* information. Often servicing information specifies a standard colour bar test signal. This type of test signal is widely used by broadcasters for checking and setting up all manner of equipment which is required to operate with colour pictures. Colour bars are also the test signal used when timing colour pictures together for use in any situation where sources need to be switched or mixed; a studio vision mixer for example. In this case the pictures need to appear in the same relative time (horizontal phase) and also the same relative colour phase (subcarrier phase) — controls normally being provided on the source equipment to vary these two parameters. Correct subcarrier phase is important because if two sources are switched, which are not in the same phase, then the monitor or video equipment will be required to relock its colour decoder, the effect will be at best, a colour flash, and at worst, complete picture break-up.

Power supply voltage range:	15 V to 25 V a.c. or d.c.
Power supply current:	105 mA at 15 V d.c.
Colour system:	PAL
Colour bar standards:	100%, EBU and 75%
Composite video output:	1 V pk-to-pk into 75 Ω (EBU bars)
UHF RF output:	591.5 MHz (channel 36)
UHF RF output connector:	phono
PCB dimensions (WDH)	
Colour bar PCB:	99 x 73 x 31 mm
Colour encoder PCB:	99 x 73 x 20 mm
Mounting holes:	M3 clear

Table 7.1   Specifications of Colour Bar Generator

Video and TV projects

Important safety note:

Because of the wide range of possible final construction methods, ultimately determined by the constructor, full details of mains wiring connections are not shown in this chapter. However for safety reasons it is essential that a suitably rated mains fuse and switch is fitted if a mains power supply is to be constructed. While by no means exhaustive, the following recommendations are made:

● if the final unit is housed in a plastic case and a mains supply is integral, Class II (double insulated) construction techniques must be employed and the mains transformer must comply with class II requirements,

● if the final unit is housed in a metal case with integral mains supply, Class I construction techniques must be employed; the case and metalwork of the mains transformer must be earthed.

Other precautions and steps necessary, to comply with published safety standards must be employed to ensure safety of the user and servicing personnel.

Every possible precaution must be taken to avoid the risk of electric shock during maintenance and use of the final unit. Safe construction of the unit is entirely dependent on the skill of the constructor.

Variations

There are several different styles of colour bars in use around the world, each suited to a particular task. They are all based on eight vertical coloured stripes, which can sometimes be seen when the vision mixer at the TV studio pushes the wrong button!

The first variation lies in the amplitude of the various parts of the waveform. The bars are actually made up of a series of black and white steps reducing in amplitude from right to left. This is the luminance (brightness) part of the signal and as a reference signal is used to set the overall gain of the signal path. 1 V pk-to-pk from the bottom of the sync pulse to the top of the white bar is the normal level. Superimposed on this luminance signal is the chrominance (colour) signal, a suppressed carrier quadrature amplitude modulated signal based on a reference frequency of 4.43361875 MHz. The amplitude and phase of this signal indicates the colour (hue) and amount of colour (saturation). There is also a quick burst of the subcarrier at reference level and phase in the line blanking interval just before the picture starts, this is the colour burst and is used to synchronise colour decoding circuitry and provide automatic level adjustment. The finer technical points of the colour encoding and decoding process have been omitted here and the reader is referred to one of the many books written on the subject.

In the old days of television the amplitude of colour bars was specified by the maximum height of the luminance signal, 1 V, and the maximum amount of colour modulation allowed, 100%. While there was only the one system of colour bars this was fine. But then it was found that for some uses 100% colour modulation was a little excessive. Setting up of radio links became difficult, and the fact that the chrominance signal extended well above the maximum white level, was also a problem So engineers began to use another set of colour bars with the amplitude reduced to 75%. The original bars were then referred to as 100% bars.

This cured the problem of too much chrominance but the white bar was also reduced to 75% and this brought its own problems. Many an engineer has seen 100% bars, thought they were 75% bars and proceeded to increase the video level accordingly, thus producing 125% pictures with disastrous results!

TV colour bar generator

To get over the problem of the 75% bars looking the same as the 100% bars, the European Broadcasting Union (EBU) designed another set of bars, known as EBU bars, having the same specifications as 75% bars except that the white bar is increased in level to 100%. The video waveforms for all three colour bar standards are shown in Figure 7.1.

With EBU standard bars the first step is bigger than the rest, but this is far outweighed by the fact that now there is a set of bars which has 100% luminance and 75% chrominance which seems to suit just about everybody.

Video and TV projects

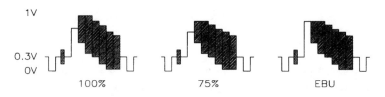

Figure 7.1 Idealised colour bar video waveforms

More variations

The second variation is how the bars are combined with other signals.

In some cases the bars are divided horizontally about two thirds of the way down the frame, normal bars appearing in the top section but other signals appearing in the bottom section. These are known as split bars and can contain various signals in their bottom section. The most common of these is a red field to the same specification as the red bar in the upper section. This is used to check for various distortions and non-linearity in the chrominance information. It also has the effect of making the dots in the red boxes on a vectorscope display larger than the rest thus making it possible to set up correct chroma phase without a colour monitor which can sometimes be useful.

A variation of the split bars has the lower section at black level with a bar inset into it at white level. This allows full 75% bars to be used in the upper section as a 100% white reference is contained in the lower section. It is also possible to observe the low frequency response of the system under test and to check for distortions such

132

as overshoot or ringing which are normally masked by the chroma information. This type of colour bars is normally found on Philips (now BTS) equipment but this is not always the case.

One exception to the normal eight bar sequence is a nine stripe design which has a white bar at the right as well as at the left. The purpose of this is to define the right-hand edge of the picture so that the picture can be accurately centred within the line blanking period when using timebase correctors or anything else where the picture is movable sideways. Due to their widespread use in video recording areas these are usually known as VTR bars.

These days it is common practice among the broadcast companies and other communications companies such as BT and satellite link providers to use split bars and insert a name or logo into the bottom section. It is easy to see how useful this is, if you think for a moment of the situation at Telecom Tower for example where incoming circuits number in the hundreds — most of them inactive most of the time and so showing colour bars as a standby signal. Just which bars are which!

Examples of all the colour bars mentioned are shown in Figure 7.2.

Colour bar generator

The colour bar generator project presented here is of the simplest configuration, that of full field bars. Any of

Video and TV projects

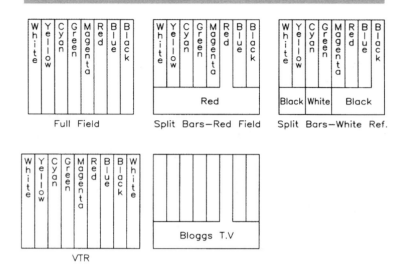

W h i t e	Y e l l o w	C y a n	G r e e n	M a g e n t a	R e d	B l u e	B l a c k

Full Field

W h i t e	Y e l l o w	C y a n	G r e e n	M a g e n t a	R e d	B l u e	B l a c k

Red

Split Bars—Red Field

W h i t e	Y e l l o w	C y a n	G r e e n	M a g e n t a	R e d	B l u e	B l a c k

| Black | White | Black |

Split Bars—White Ref.

W h i t e	Y e l l o w	C y a n	G r e e n	M a g e n t a	R e d	B l u e	B l a c k	W h i t e

VTR

Bloggs T.V

Figure 7.2   Various colour bar screen displays

the three 100%, EBU, or 75% standard levels can be generated — the choice is either set by jumpers or by an off-board switch. The project is based on two PCBs: the first contains power supply circuitry and generates the necessary video and timing signals — this is referred to as the colour bar module. The second encodes the colour video signals to the PAL TV standard and provides composite video and modulated UHF RF outputs — referred to as the colour encoder module.

To help understand operation of the colour bar generator Figure 7.3 shows a block diagram of the overall project.

134

Circuit descriptions

Colour bar module

The circuit of the colour bar module is shown in Figure 7.4. Either an a.c. or d.c. extra low voltage supply can be applied to the power connector PL1; in the case of an a.c. supply D1 and D2 form a bi-phase full wave rectifier, in the case of a d.c. supply the same diodes provide reverse polarity protection. C1 is the main reservoir capacitor. Since the equivalent series resistance (ESR) of large value electrolytic capacitors increases significantly at high frequencies, C2 provides high frequency decoupling thus attenuating supply borne noise and spikes. The incoming supply is regulated to +12 V d.c. by RG1. C3 provides high frequency decoupling at the output of the regulator to promote stability. LD1 pro-

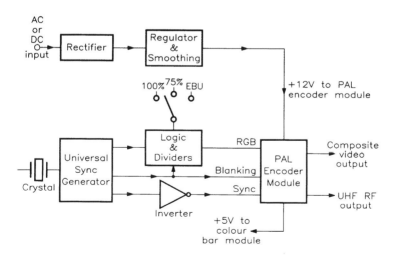

Figure 7.3   Block diagram of Colour Bar Generator

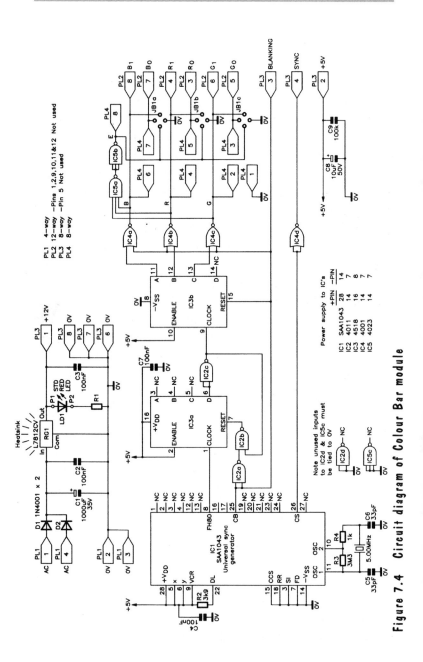

Figure 7.4 Circuit diagram of Colour Bar module

vides power on indication; R1 limits the current through LD1 to approximately 20 mA. The regulated +12 V supply is required by the colour encoder module and supplied to it through PL3–1, which in turn supplies a +5 V regulated supply back to the colour bar module through PL3–2 — a symbiotic existence — each module requires the other to operate. The +5 V supply from the colour encoder module is decoupled by C8 and C9. Additional decoupling is provided by C4 (physically adjacent to IC1) and C7 (physically adjacent to IC2 to IC5). Such decoupling is essential when working with digital logic ICs to prevent erratic operation, digital noise breakthrough and reduce electromagnetic interference (EMI).

The heart of the colour bar module is IC1 which is an SAA1043 Universal Sync Generator IC, this device generates all the required timing signals for a television picture. Before such devices were developed, a whole board of logic ICs would be necessary to generate the required timing signals. The IC does not, however, generate the colour subcarrier which is generated by the colour encoder module. IC1 has several programming pins which define its operating parameters such as the TV standard it is to be used on. These are tied either high or low as shown in the circuit diagram.

Very few external components are required for the SAA1043 to operate. Crystal XT1, its associated RC network comprising R3 and C5, and, R4 and C6, and circuitry within the IC form an accurate 5 MHz oscillator. This oscillator provides the master reference for all system timing. The only other component is a single pull up resistor on the DL input.

Video and TV projects

One problem often encountered when generating timing signals electronically is finding a locked source of correctly timed pulses at a much higher frequency than the normal line and frame rate, so that the picture area can be split into various vertical segments. Fortunately the SAA1043 has an output of 1.25 MHz which is eighty times the line rate and is, of course, locked to it.

The FH80 (eighty times the line rate) output signal is passed to the clock input of IC3A, a 4518 dual BCD counter. The D output from IC3A is now FH80 divided by eight, which gives ten times the line rate (FH10). This would give ten bars, not eight, but most of the unwanted two bars are contained within the line blanking period and as such are gated off by the $\overline{CB}$ (negative going composite blanking) signal from IC2A. IC2A (used as an inverter) provides CB from the CB output of IC1. This is NANDed with the $\overline{D}$ output from IC2C (also used as an inverter), to reset IC3A. The input to IC3B is a series of eight normal width pulses and a final narrow pulse giving nine in total. From this it will be seen that the colour bars produced will have nine bars, not eight, the final right hand bar being a narrow white bar. This will actually give a form of VTR bars; for most, if not all, applications this will be an advantage not a disadvantage. The CB output signal from IC1 is fed to the colour encoder module's blanking input through PL3–3.

The colour encoder module requires RGB (Red, Green and Blue) component video signals. The combinations and relative timing of RGB signals to produce familiar staircase luminance waveform are shown in Figure 7.5. The encoder sums the RGB signals at the correct levels to generate the luminance signal, superimposes the chrominance information and colour burst, and adds the synchronising signals resulting in the waveforms shown in Figure 7.1.

To provide the RGB signals, the FH10 signal from IC2C needs to be divided by two, four and eight to give $\overline{B}$, $\overline{R}$ and $\overline{G}$ signals respectively. This is achieved by IC3B. IC4A to IC4C invert the $\overline{RGB}$ signals to provide RGB signals for the inputs of the colour encoder module. These three gates, by means of the CB signal fed to the second input of each gate, remove any signals outside the active picture area.

IC5 detects when all the three bar outputs are high, which is only during the white bar period, and generates a peak white bar signal, sometimes called the E bar. This is added to the three colour outputs in the colour encoder module to give a true EBU level output. Figure 7.5 indicates how the E bar is used to achieve this.

The inputs to the colour encoder module are in the form of a two bit per colour TTL compatible port. The RGB signals are input into the MSB inputs (Rl, G1 and Bl). What is input into the LSB inputs (R0, G0 and B0) determines the final form of the Colour Bar output. Connecting all three LSB bits to ground will give 75% bars with a 75% white reference. Connecting all three LSB bits to the output from IC5 (E signal) will increase the level during the white bar giving 75% colour and 100% white reference, that is to say EBU bars. Connecting each LSB to its associated MSB will produce 100% bars with 100% white reference. The RGB signals are to the colour encoder module through PL2–3 to PL2–8.

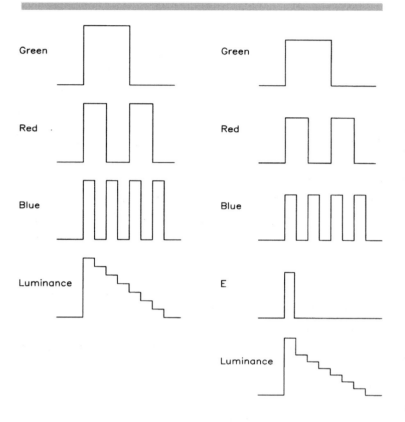

Figure 7.5 Relative timing of component video signals

Selection of the required colour bar standard is achieved by a jumper block, JB1a to JB1c, on the PCB or a three pole switch connected to PL4.

The CS (composite sync) output signal from IC1 is inverted by IC4D to give $\overline{CS}$ (negative going composite sync) which is fed to the colour encoder module's sync input through PL3–4.

TV colour bar generator

Colour encoder module

The colour encoder module is shown in block diagram form in Figure 7.6; the circuit diagram is shown in Figure 7.7.

The +12 V supply for the colour encoder module is provided by the colour bar module as previously explained and is supplied through connector PL4–1. The +12 V supply feeds the video buffer and the TEA2000 colour encoder IC. C1 provides supply decoupling, additional high frequency decoupling is provided by C3 and C12. The RF Modulaton MD1, requires a +5 V supply which is provided by RGl, the output of which is decoupled by C4 and C5. The +5 V supply is also fed back to the colour bar module through PL4–2.

IC1, a TEA2000, is a colour encoder and video summer which has an internal oscillator from which R–Y and B–Y colour difference signals are produced. As can be seen from the block diagram shown in Figure 7.6, the chip has a complex internal structure and requires very few additional components. The frequency of the internal oscillator is set by XT1 an 8.867238 MHz crystal (twice the colour subcarrier frequency). The output of the IC's oscillator stage is divided to provide the four subcarrier phases required in the encoder.

The combined luminance and sync signal appearing at pin 7 of IC1 is coupled to pin 8 via a 270 ns delay line.

Chrominance filtering is accomplished with a parallel tuned circuit comprising C8 and L2 a.c. coupled to pin 10 of IC1 by C7. The filter is set to resonate at 4.433 MHz.

Video and TV projects

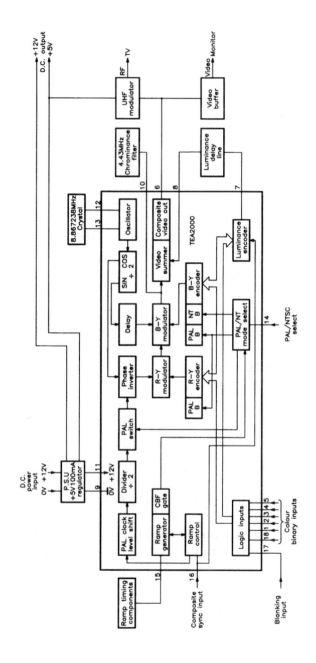

Figure 7.6 Block diagram of Colour Encoder module

142

TV colour bar generator

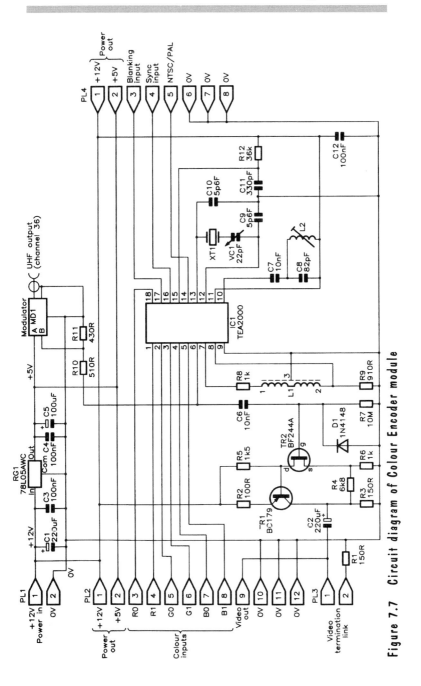

Figure 7.7 Circuit diagram of Colour Encoder module

Video and TV projects

The timing information for the colour burst and PAL switching is controlled by the ramp timing components, C11 and R12.

The composite video signal from pin 6 of IC1 is supplied to the UHF modulator, MD1, by a potential divider formed by R10 and R11. This signal is also capacitively coupled, by C6, to the input of the video buffer.

The input impedance at the gate of TR2 is approximately 10 MΩ, whilst the output impedance of TR1 is much lower, approximately 100 Ω. The buffered video signal is supplied by C2 to the video output on PL2–9. When using a video monitor with a high input impedance the termination resistor, R1, must be placed in circuit by linking PL3–1 and PL3–2.

Construction

Assembly of the two PCBs should prove straightforward providing a logical assembly sequence is followed. Double check component type, value and polarity before soldering as subsequent component removal may damage PCB tracks unless great care is taken. It is recommended that the smallest components are fitted first (wire links, diodes, resistors, etc.), working up to the largest components (large electrolytic capacitors, modulator, regulator and heatsink, etc.). It is a good idea to leave fitting the ICs into their sockets until last; precautions should be taken to prevent electrostatic discharge as this may permanently damage the ICs or cause premature failure in service. For further information on general construction techniques and component identification, please refer to the constructors' guide which is included in the kits. Figures 7.8 and 7.9 show the PCB legend and track for the colour bar module and the colour encoder module respectively. After the PCBs have been assembled, remove excess flux from the board using an environmentally friendly PCB cleaner and double check for misplaced components, solder splashes, etc.

The PCBs are interconnected by means of ribbon cable and Minicon connectors. Connections should be made as indicated in the wiring diagram shown in Figure 7.10. Selection of the colour bar standard is by means of PCB jumpers or an offboard switch. Jumper positions and wiring for the switch are also shown in Figure 7.10. Do not fit more than one jumper per group. If the off-board switch is to be used, do not fit any jumpers.

Figure 7.11 shows three possible power supply options; these are:

● a mains transformer with a centre-tapped 12–0–12 V to 15–0–15 V secondary winding and able to supply at least 150 mA,

● a mains transformer with twin 12 V to 15 V secondary windings and able to supply at least 150 mA,

● a power supply with an output voltage between +12 V and +25 V d.c. and able to supply at least 150 mA.

Since the PCBs have the same dimensions and fixing centres, they may be piggy-back mounted as shown in Figure 7.12.

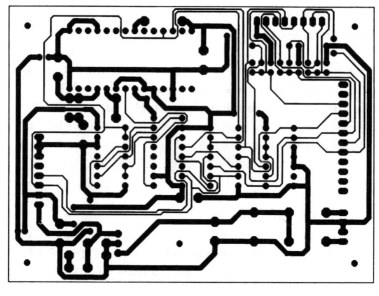

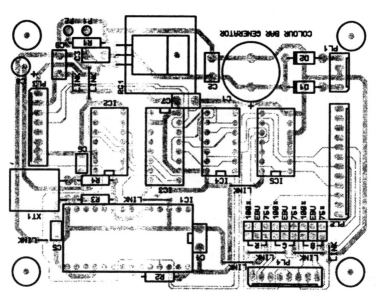

Figure 7.8 PCB legend and track for Colour Bar module

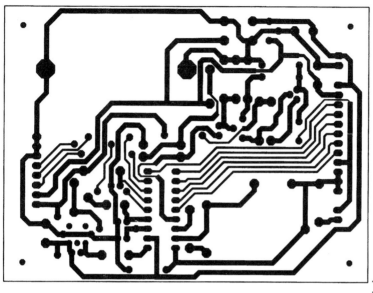

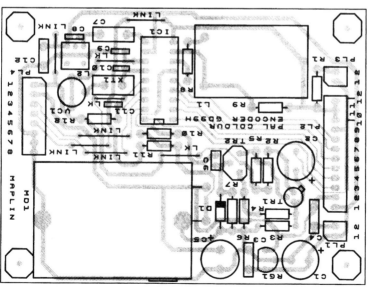

Figure 7.9 PCB legend and track for Colour Encoder module

Video and TV projects

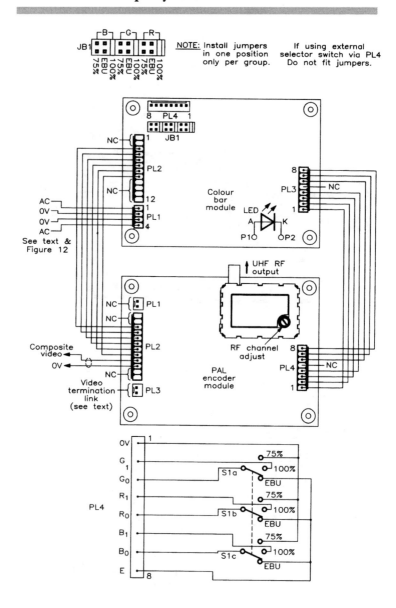

Figure 7.10 Wiring diagram and jumper selection

Testing and alignment

For the purposes of testing and alignment a power supply able to supply between +15 V and +25 V d.c. is
required.

Connect the supply to the colour bar generator as shown
in Figure 7.11 with a multimeter set to read d.c. mA on a
250 mA or higher range in series with the positive supply. Measure the current, which should be approximately
100 mA.

Remove the multimeter from the supply, set it to read
d.c. V on a 15 V or higher range and reconnect the sup-

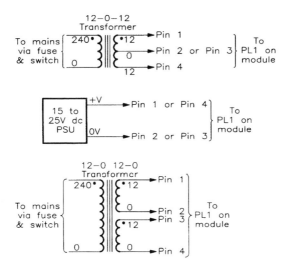

Figure 7.11   Power supply options

Video and TV projects

ply. On the colour bar module measure the voltage on PL3–1 with respect to PL3–8, which should be approximately +12 V. Measure the voltage on PL3–2 with respect to PL3–8, which should be approximately +5 V. If readings are markedly different from those listed, disconnect the supply and recheck for errors in construction.

Connect the composite video output to a colour monitor or the UHF RF output to a colour TV (TV will require

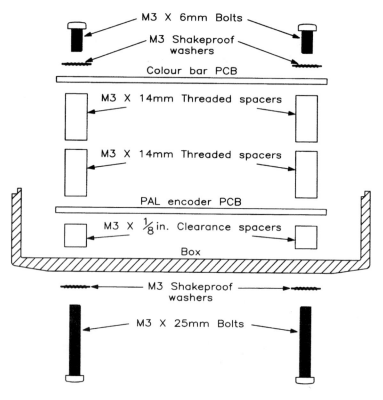

Figure 7.12 Piggyback mounting arrangement

tuning). Once a picture is displayed, adjust VC1 on the colour encoder PCB for correct locked colour. Using an oscilloscope monitor the composite video output and adjust L2 for minimum rounding or overshoot on the chrominance envelope of the video waveform. If an oscilloscope is not available, adjust L2 for best picture, i.e. minimum colour bleed between adjacent bars and minimum colour variation between adjacent TV lines (venetian blind effect).

The output frequency of the RF modulator is factory set to channel 36. If it is necessary to adjust the frequency, this can be achieved by retuning the ferrite core visible from the top of MD1's case. Since the ferrite core is extremely fragile a suitable non-magnetic adjustment tool must be used — do not use a screwdriver. Adjustment of the modulator frequency may be necessary, e.g. when adding the RF signal into a TV distribution amplifier in a TV repair workshop to avoid clashing with other equipment on channel 36, e.g. satellite receiver, VCR, etc.

Professional users please note

Although the colour bar generator appears to generate bars to full EBU specification this is in fact not the case. The line and frame frequency rate, while adequate for viewing on a monitor, will not be sufficiently accurate to record on a professional VTR (although a VHS recorder should be fine). For the same reason the colour bar generator should not be used as a Genlock source in multicamera studios. The EBU specification includes a requirement for the subcarrier to be locked to the horizontal timing pulses. Of course this is also not the case in the colour bar generator as separate crystals are used to obtain the luminance and chrominance parts of the output.

Colour bar module parts list

Resistors — All 0.6 W 1% metal film

R1	560 Ω	1 (M560R)
R2	3k9	1 (M3K9)
R3	3M3	1 (M3M3)
R4	1 kΩ	1 (M1K)

Capacitors

C1	1,000 µF 35 V PC elect	1 (FF18U)
C2,3,4, 7,9	100 nF monolithic ceramic	5 (RA49D)
C5,6	33 pF ceramic	2 (WX50E)
C8	10 µF 50 V PC elect	1 (FF04E)

Semiconductors

D1,2	1N4001	2 (QL73Q)
RG1	L7812CV	1 (QL32K)
LD1	5 mm LED red	1 (WL27E)
IC1	SAA1043	1 (UK85G)
IC2	4011 BE	1 (QX05F)
IC3	4518BE	1 (QX32K)
IC4	4001BE	1 (QX01B)
IC5	4023BE	1 (QX12N)

Miscellaneous

XT1	5 MHz crystal	1 (UL51F)
	28-pin DIL IC socket	1 (BL21X)

Video and TV projects

	16-pin DIL IC socket	1 (BL19V)
	14-pin DIL IC socket	3 (BL18U)
PL1	4-way minicon plug	1(YW11M)
PL2	12-way minicon plug	1(YW14Q)
PL3,4	8-way minicon plug	2 (YW13P)
	4-way minicon socket	1 (HB58N)
	12-way minicon socket	1 (YW24B)
	8-way minicon socket	2 (YW23A)
	minicon terminal strip	3 (YW25C)
JB1	2 x 36 pin strip	1 (JW62S)
	mini pin jumper	3 (UL70M)
	1 mm veropin	1 (FL24B)
	slotted heatsink	1 (FL58N)
	M3 x 10 mm bolt	1 (JY22Y)
	M3 shakeproof washer	1 (BF44X)
	M3 nut	1 (JD61R)
	PCB	1 (GH67X)
	instruction leaflet	1 (XU60Q)
	constructors' guide	1 (XH79L)

Optional (not in kit)

3-pole 4-way rotary switch	1 (FF75S)
10-way ribbon cable	1 (XR06G)
miniature coax	as req (XR88V)
1.4 A wire red	as req (BL07H)
1.4 A wire black	as req (BL00A)
phono to coax cable	1 (FV90X)
M3 x 25 mm bolt	1 (JY26D)
M3 x 6 mm bolt	1 (JY21X)
M3 nut	1 (JD61R)
M3 shakeproof washer	1 (BF44X)
M3 x 14 mm threaded spacer	1 (FG38R)
M3 x 1/8 in clearance spacer	1 (FG32K)

Pal colour encoder parts list

Resistors — All 06 W 1% metal film

Rl,3	150 Ω	2	(M150R)
R2	100 Ω	1	(M100R)
R4	6k8	1	(MK6K8)
R5	1k5	1	(M1K5)
R6,8	1 k	2	(M1K)
R7	10 M	1	(M10M)
R9	910 Ω	1	(M910R)
R10	510 Ω	1	(M510R)
R11	430 Ω	1	(M430R)
R12	36 k	1	(M36K)

Capacitors

C1,2,5	22 Ω F 16 V PC electrolytic	3	(FF13P)
C3,4,12	100 nF minidisc	3	(YR75S)
C6	10 nF minidisc	1	(YR73Q)
C7	10 nF poly layer	1	(WW29G)
C8	82 pF ceramic	1	(WX55K)
C9,10	5p6F ceramic	2	(WX41U)
C11	330 pF ceramic	1	(WX62S)
VC1	22 pF trimmer	1	(WL70M)

Semiconductors

IC1	TEA2000	1	(UH66W)
RG1	μA78L5AWC	1	(QL26D)
TR1	BC179	1	(QB54J)
TR2	BF244D	1	(QF16S)
D1	1N4148	1	(QL80B)

Video and TV projects

Miscellaneous

L1	delay line	1	(UH84F)
L2	15 µH adjustable coil	1	(UH86T)
XT1	8.867238 MHz crystal	1	(UH85G)
MD1	UM1233 UHF modulator	1	(FT30H)
PL1,3	2-way minicon latch plug	2	(RK65V)
PL2	12-way minicon latch plug	1	(YW14Q)
PL4	8-way minicon latch plug	1	(YW13P)
	18-pin DIL IC socket	1	(HQ76H)
	2-way minicon latch housing	2	(HB59P)
	12-way minicon latch housing	1	(YW24B)
	8-way minicon latch housing	1	(YW23A)
	minicon terminal	3	(YW25C)
	PCB	1	(GD99H)

Optional (not in kit)

trim tool	1	(BR51F)
video lead 6	1	(FV90X)
M3 x 14 mm threaded spacer	1	(FG38R)
M3 x 6 mm bolt	1	(BF51F)
M3 shakeproof washer	1	(BF44X)
M3 nut	1	(BF58N)

8 Video box

It's a relatively simple task to fade out an audio signal. It's achieved by reducing the amplitude of all its frequency components at the same time. In its most basic form this can be accomplished by using nothing more than a potentiometer, tapping off the required amount of audio signal. However, composite video signals are made up from several different elements (colour, brightness and timing), and if all are reduced by the same amount at the same time, the picture will begin to break up long before maximum fade to black is reached.

To maintain a stable picture, a more complex signal processing arrangement is required. The Video Box, shown in block form in Figure 8.1, permits the reduction in amplitude of picture information, while maintaining the level of the timing signals used to synchronise it. To

Video and TV projects

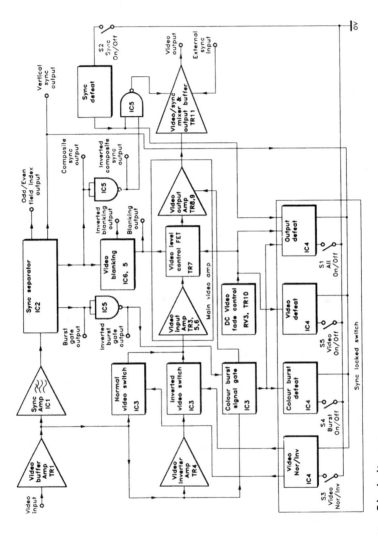

Figure 8.1 Block diagram

achieve this, the incoming composite video signal must be split up into its component parts, each of which undergo selective processing. Following this, these components are reassembled, to produce the final composite output. An additional video inverter has been included, allowing the option for a negative picture to be displayed (useful for home video and long distance/ satellite TV enthusiasts!). Apart from its own function as a video fader, the system timing signals have been made available, and as a result the finished unit can be used as a versatile building block for use within other video projects.

Circuit description

In addition to the block diagram shown in Figure 8.1, a detailed circuit diagram is provided in Figure 8.2. These two diagrams should assist you when following the circuit description, or fault-finding in the completed unit. For the circuit to function correctly, it must be powered from a well-regulated +12 V d.c. supply. This supply enters the unit on PL6, and must have the correct polarity (negative on pin 1 and positive on pin 2), otherwise damage may occur to the semiconductors. The main supply decoupling is provided by C2, 15, 19 and C22; with C1, 14, 18, 21 and C27 giving additional high-frequency suppression.

The incoming video is connected to PL1 (pin 1 ground, pin 2 signal), and is terminated by R1. The video signals are a.c.-coupled via C3 into the video buffer, an emitter-follower stage based around TR1. This signal is fed to the following video processing circuits:

Video and TV projects

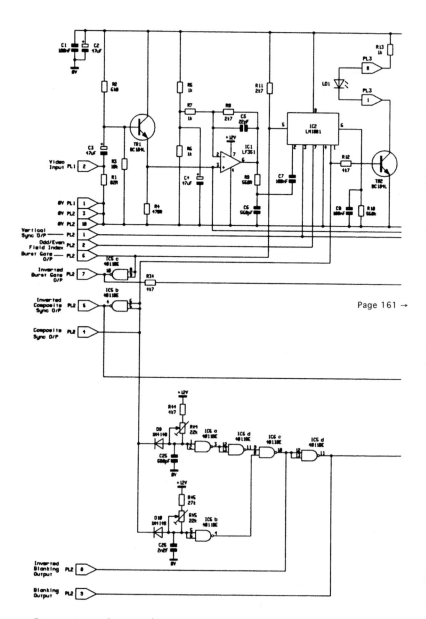

Page 161 →

Figure 8.2   Circuit diagram

160

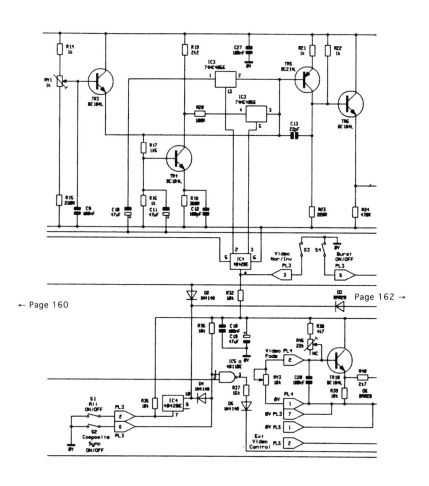

← Page 160

Page 162 →

Figure 8.2   Continued

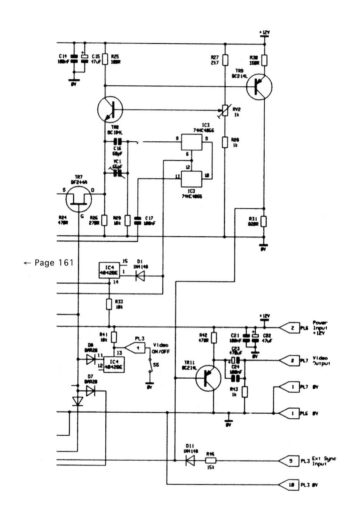

← Page 161

Figure 8.2 Continued

162

- sync amplifier IC1,
- normal video switch IC3,
- video inverter amplifier TR4,
- colour burst signal gate IC3.

The function of IC1, the sync amplifier, is to obtain the maximum amount of sync information from the composite video signal. This is brought about by the circuit's combination of gain, d.c. input bias and frequency response. The cleaned-up and filtered signal is then fed via C7 to the input (pin 2) of the sync separator, IC2. This LM1881 device is a dedicated chip that extracts the timing information (sync) from the processed composite video signal. Four major sync signals are produced by the IC:

- composite sync (on pin 1),
- vertical sync (on pin 3),
- colour burst gate (on pin 5),
- odd/even field index (on pin 7).

These timing signals are at +12 V CMOS logic levels, and are used by some of the other stages within the Video Box system to synchronise events. They are also available (some with inverted logic) on PL2. Here, the inverted composite sync is provided by IC5b, while the inverted colour burst gate is supplied by IC5c. The inverted composite sync from pin 4 of IC5b passes through an inverting gate, IC5a, before being mixed back with the picture information at TR11. This gate is used to turn off, or remove, the composite sync from the final composite video output, which appears at pin 2 of PL7. This function is activated when S1 and/or S2 are closed.

Video and TV projects

The composite sync signal is also used by the input detector (TR2) and video blanking (IC6) circuits. With no signal present at the video input (PL1 pin 2), no sync pulses will be produced by IC2. Under this condition, TR2 will not conduct, so indicator LD1 will not light up — until a signal is available. From the composite sync, the video blanking circuit generates a timing signal, which is used by the main video amplifier to ensure that only picture information is processed. This blanking signal is also available on PL2 pin 9, with its inverted logic condition present on pin 8. The fine tuning of this event is controlled by RV4, which is used to set the start (or left edge), while the end (or right edge) of the picture line is set by RV5.

The vertical sync signal from pin 3 of IC2 is used to control the sync-locked switch functions; video normal/inverse, video defeat, colour defeat and output defeat. This signal is also available on pin 1 of PL2. By locking these switches to the vertical sync, a cleaner and more professional picture change is obtained.

This is achieved by using a quad D-type latch IC4 (4042BE), which has its clock input (pin 5) connected to the vertical sync signal. Only when this signal is present will the output of this IC change state upon operating one of the above function switches, the sync-locked logic outputs from this IC being used to control the system's sync, colour-burst and video processing circuits.

The colour-burst gating signal from pin 5 of IC2 is connected to pin 6 of PL2, and the input of IC5c (which is used as an inverter). The output of this inverter is connected to pin 7 of PL2, and also passes via R34 to the colour-burst signal gate IC3. However, before it reaches its final destination it can be interrupted by being pulled down to ground by the colour-burst and/or output defeat switches.

The odd/even field index signal from pin 7 of IC2 is not used by the Video Box. However, it is available on pin 2 of PL2 for possible use in additional circuitry. For example, this output could be useful in frame memory storage applications, or in extracting test signals that may occur only in alternate fields.

There are two paths that the video signal can take on its way to the main video processing amplifier, both of which pass through IC3. This analogue switching device is used to select the normal or inverted (picture negative) video information, and is controlled by two of the logic outputs from IC4 (S3). When the normal (or non-inverted) signal is selected, the signal path is from the output of the video buffer amp, TR1, to the input of the main amp, TR5. When the inverted signal is selected, the video path is from the output of the video inverter amp, TR4, to the input of the main amp, TR5. The d.c. bias for both TR4 and TR5 is provided by TR3, and is set by RV1.

The colour-burst signal must remain at a constant level as the video picture information is reduced. Otherwise, the monitor/VCR will try to over-compensate by increasing the colour intensity as the video level is reduced. Eventually, its colour-killer circuit will operate. At the point just before this threshold is reached, the colour video information will appear very noisy (when viewed on a monitor) before suddenly cutting out, leaving a black-and-white (monochrome) picture on the screen. The colour-killer circuit is included by the majority of video/TV equipment designers to accommodate the possibility of a monochrome signal, or if the overall signal is too weak/noisy to reproduce a good enough quality colour picture.

The colour-burst is maintained by gating the composite video signal from TR1 at a precise moment in time, and then recombining it after the video level control circuit, TR7. This signal gating is achieved by IC3, which receives its switching control signal from the logic output of IC5c. However, it can be removed if the colour-burst defeat switch, S4, and/or output defeat switch, S1, is activated.

The main video level processing amplifier consists of four stages:

● the d.c. bias for the input buffer is supplied by TR3. The output of this stage is determined by the level of voltage applied to the base of TR3 by RV1,

● the input buffer, formed by TR5 and TR6, then conditions the video signal into a form suitable for feeding TR7, the video level control FET,

● TR7 is controlled by the d.c. voltage applied to its gate; the higher this is, the higher the amplifier gain will be. This voltage is supplied by the d.c. video fade control transistor, TR10. A potential divider on the base of TR10 is created by using three resistors: a fixed resistor, R38; a preset, RV6; and a rotary (or slide) control, RV3. The combined value of R38 and RV6 sets the maximum voltage limit, which corresponds to the maximum video level. However, as the value of RV3 decreases, this voltage will drop until the zero point is reached, corresponding to minimum video level. The voltage output from TR10 is grounded by the video blanking circuit, which is based around IC5d and IC6. This has the effect of turning off the main video amplifier during the sync and colourburst time slots. The voltage can also be manually grounded if the video defeat (S5) and/or output defeat (S1) switches are activated. Additionally, it can be grounded by an external influence applied to pin 2 of PL5.

● the output amplifier, comprised of TR8 and TR9, receives the video signals from TR7, and the colour-burst signal from IC3. Two capacitors, C16 and VC1, are used to set the injection level of the colour burst. A second d.c. bias control, RV2, sets the black level of the video signal.

The output from the main video processing amplifier is supplied to TR11, where the previously extracted composite sync is mixed with it to produce the final composite colour video output signal. As well as mixing the two signals, TR11 acts as a buffer, providing a low impedance 75 R output drive which passes through C23 and C24, onto pin 2 of PL7. This video/sync mixer stage can also be supplied with external sync information applied to pin 9 of PL3.

PCB Assembly

All the information required to help you with soldering and assembly techniques, should you need it, can be found in the Constructors' Guide included in the kit. The printed circuit board (PCB) is a single-sided glass fibre type, chosen for maximum reliability and mechanical stability. Removal of an incorrectly-fitted component can be fairly difficult without damaging it, or the PCB, in some way, so please double-check each component type, value, and polarity (where appropriate), before soldering! The PCB has a legend to assist you in correctly positioning each item (see Figure 8.3). It is usually easier to start with the smaller components, such as the resistors. Next, mount the ceramic, polylayer, polystyrene and electrolytic capacitors. All the diodes have a band at one end; be sure to position them according to the legend. When installing the transistors, make certain that each case matches its outline. This also applies to the IC sockets and Minicon connectors, where you should match the notch with the block on the legend. The next components to be installed are the five preset resistors (RV1, 2, 4, 5, 6) and the trimmer capacitor, VC1. Only after all the other components have been fitted should you then carefully insert the relevant ICs into their sockets, making sure that you correctly orientate them. Finally, do not forget to fit all ten wire links. These can be made from component lead off-cuts, or the 22SWG tinned copper wire included in the kit. Photo 8.1 shows in detail the completed PCB assembly.

This completes the assembly of the PCB, and you should now check your work very carefully, making sure that all the solder joints are sound. It is also very important to ascertain that the solder side of the circuit board does not have any trimmed component leads protruding by more than 2 mm, as a short-circuit may otherwise result.

Final assembly

No specific box has been designated for the project, as your finished unit could contain other PCBs (e.g. video digitiser and computer cards). However, the basic unit fitted nicely in to an ABS console M6006 (stock code LH66W). This, and the additional connectors and hardware, are listed under *Optional (not in kit)*, in the parts list. Once you have completed the mechanical assembly of the unit, you should check your work very carefully before proceeding to the wiring stage.

Wiring

If you purchase the Maplin kit (stock code LP48C), it should contain a one metre length of ten-way ribbon cable. This is used for all the d.c. connections. However, no specific colour has been designated for each wire connection; this choice is left entirely to you. Coloured wire is used to simplify matters, making it easier to trace separate connections to off-board components, just in case there is a fault in any given part of the circuit. Miniature 75 Ω coaxial cable is used for the video in/out signals on PL1 and PL7, and it is most important that the braided screen should not be able to come into contact with the centre conductor, or anything connected to it. All the wire connections to the PCB are made using *Minicon* connectors; the method of installing these is shown in Figure 8.4. A wiring diagram, which shows all of the interconnections on the PCB, is given in Figure

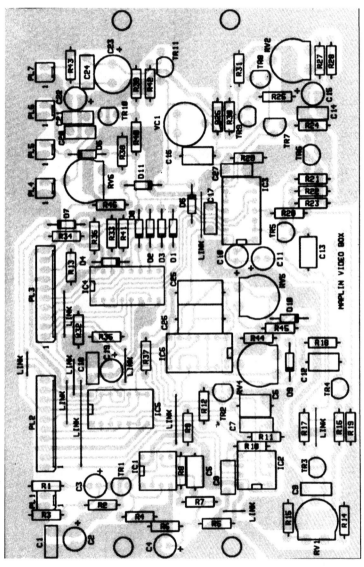

Figure 8.3 PCB legend and track

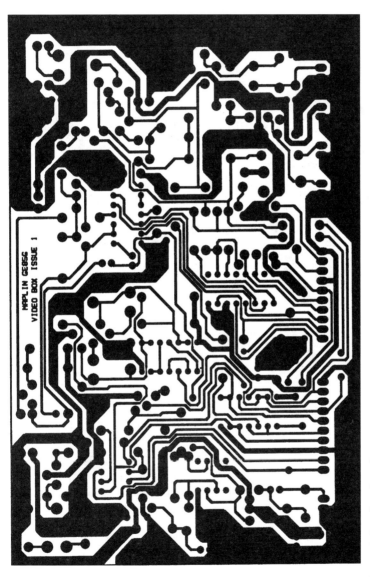

Figure 8.3 Continued

Photo 8.1 Completed PCB

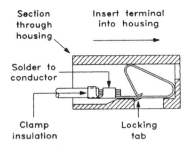

Figure 8.4   Terminating the wires

8.5. The actual physical connections to the rotary video fade potentiometer, and the optional slide control, are shown in Figure 8.6.

This completes the wiring of the Video Box and you should now check your work very carefully, making sure that all the solder joints are sound.

Testing and alignment

The initial d.c. testing procedure can be undertaken using the minimum amount of equipment. You will need a multimeter and a well-regulated +12 V d.c. power supply, capable of providing at least 150 mA. All of the following readings are taken from the prototype using a digital multimeter, and some of the readings you obtain may vary slightly, depending upon the type of meter used!

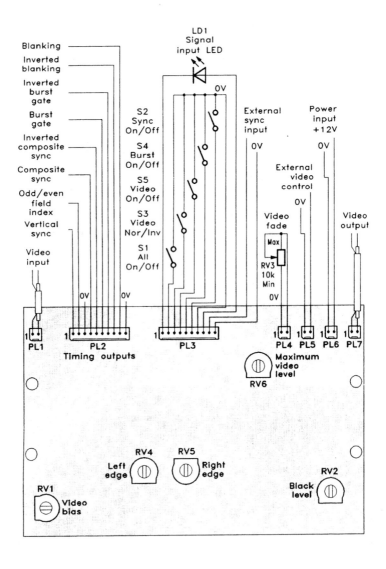

Figure 8.5   Wiring

Before you commence testing the unit, set the PCB presets (RV1, 2, 4, 5, 6 and VC1) to their half-way positions, and the off-board function switches (S1, 2, 3 and 4) to their *open* or *off* positions. Ensure that the rotary (or slide) video fade control, RV3, is set to its maximum level, as shown in Figure 8.6. *Do not* connect any power or video signals to the unit at this stage of testing.

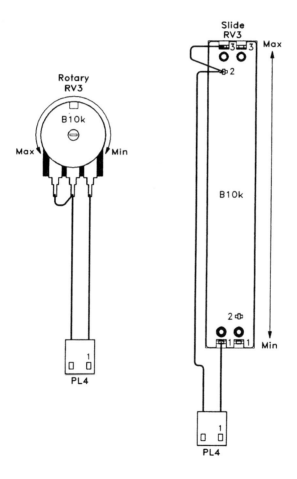

Figure 8.6 Video fade potentiometers

Video and TV projects

The first test is to ensure that there are no short-circuits before connecting the unit to a d.c. power supply. Set your multimeter to read *ohms* on its resistance range, and connect the two test probes to pin 1 and pin 2 of PL6. With the probes either way round, a reading greater than 60 Ω should be obtained. If a much lower reading is registered, then check that solder joints and component leads are not shorting between tracks. Next, monitor the supply current; set your meter to the d.c. mA range and place it in series with the positive line, pin 2 of PL6. Connect up and switch on your regulated +12 V power supply, ensuring correct polarity. A current reading of approximately 110 mA should be obtained, and the signal input indicator LD1 should not be illuminated. Switch off, disconnect the power supply and remove the meter.

Reconnect the power supply, and set your multimeter to read d.c. volts. All of the voltages are positive with respect to ground, so connect your negative test lead to any of the 0 V ground pins on the terminal blocks:

PL1 pin 1	PL5 pin 1
PL2 pin 3 or pin 10	PL6 pin 1
PL3 pin 7 or pin 10	PL7 pin 1
PL4 pin 1	

The voltages present on the PCB assembly should approximately match the readings shown in Table 8.1. When these tests have been completed successfully, remove power from the unit.

To commence the video testing and alignment, you will require some additional video equipment and test gear. You will need a source of composite colour video which

could be a VCR, TV tuner, video camera or pattern generator. To observe the resultant output, you will require
a composite colour video monitor (or a TV/VCR with a
video input) and a 20 MHz oscilloscope with TV coupling.
If you haven't got access to the more exotic video test
gear, *don't panic!*— it is still possible to obtain acceptable results using only the video equipment and your
own visual judgement. The picture image used in our
tests was taken from a colour video camera, and is of an
old Maplin catalogue.

D.C. power requirement:	+12 V
D.C. current:	120 mA
Video system:	composite PAL colour video;
	1 V peak-to-peak
Video gain:	0 dB
Frequency bandwidth:	10 MHz
Input impedance:	75 Ω (nominal)
Output impedance:	75 Ω (nominal)
Video control:	fade to black
	external control
Synchronized switches:	video normal or inverted
	video defeat
	colour burst defeat
	output defeat
Composite sync:	on/off
	external input
Timing outputs:	composite sync
(all +12 V CMOS)	inverted composite sync
	vertical sync
	odd/even field index
	video blanking
	inverted video blanking
	colour burst gate
	inverted colour burst gate

Table 8.1   Specification of prototype

Video and TV projects

Connect the video source to the input (PL1), and the monitor to the output (PL7). If you have an oscilloscope, set it up as follows: TV coupling; horizontal time-base 10 μs; auto-trigger with negative polarity; vertical input 0.2 V/div d.c. coupled. Next, connect its vertical input to the video output on PL7.

Ensure that all of the function switches on the Video Box are set to their *off* or *open* positions, and that the video fade control, RV3, is set to maximum. Apply power to the unit and, if all is correct, LD1 should light and you should observe a clear high-quality picture on the monitor screen (see Photo 8.3) and a composite video waveform on the oscilloscope display (see Photo 8.2). However, this is unlikely since none of the presets have been critically adjusted, and so the picture is more likely to resemble the one shown in Photos 8.4 and 8.5. Each preset will affect a particular parameter of the video signal. Interaction may occur between several of these adjustments, so progressive recalibration may be required.

The first adjustment is to set the correct input and black-level bias points. As RV1 is turned anticlockwise, the picture quality will progressively deteriorate as the level of high-frequency distortion increases. This will continue until the picture inverts on highlights, as shown in Photo 8.19. Turn RV1 clockwise until this effect is corrected; over adjustment of this preset will have an adverse result on the inverse-video mode. The black-level is set by RV2, and as this is turned anticlockwise the video information will progressively drop down into the sync region, eventually causing picture break-up (see Photos 8.6 and 8.7). If this preset is turned too far clockwise, the picture will look washed out (see Photos 8.8 and 8.9). RV1 and RV2 do have a small effect on each other, and so progressive readjustment may be necessary.

The next parameter to correct is the amount of picture information shown on the screen. The left-hand edge is adjusted using RV4, while the right-hand edge is set by RV5 (see Photos 8.10 to 8.13 inclusive).

If an oscilloscope is not available, RV6 and VC1 will not be as easy to set up as the other presets. This is because RV6 sets the maximum video level, and most modern equipment can accept a significantly over-driven input, particularly VCRs, which feature a video AGC circuit. The colour-burst signal level, set by VC1, is used to control the colour intensity of the picture. To calibrate the unit to give a standard 1 V peak-to-peak video output, an oscilloscope will be required (see Photo 8.2). However, a rough setting can be made by comparing the relative brightness and colour intensity of the direct video signal to the processed one.

Finally, test the following functions:

● range-test the video fade control, RV3. The control should give complete control from maximum picture level, down to a black screen (as shown in Photo 8.18),

● output defeat switch, S1. All video signal components should be blocked when S1 is closed,

● sync defeat switch, S2. The composite sync pulses should be removed when S2 is closed, resulting in a slipping picture, as shown in Photos 8.14 and 8.15.

● video invert switch, S3. Picture should become a negative image when S3 is closed — see Photo 8.20,

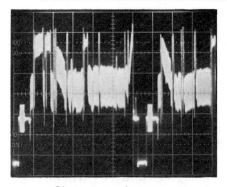

Photo 8.2   Correct composite
video waveform (CRT display)

Photo 8.3   Correctly aligned picture
(monitor screen)

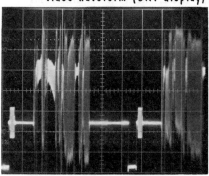

Photo 8.4   Incorrectly
aligned picture (CRT display)

Photo 8.5   Incorrectly aligned
picture (monitor screen)

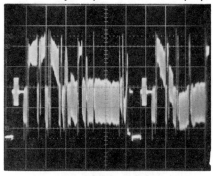

Photo 8.6   Incorrectly
aligned black-level — too
low (CRT display)

Photo 8.7   Incorrectly aligned black
level — too low (monitor screen)

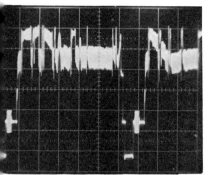

Photo 8.8   Incorrectly
aligned black-level — too
high (CRT display)

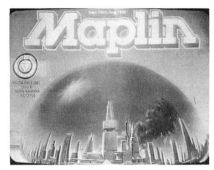

Photo 8.9   Incorrectly aligned black-
level — too high (monitor screen)

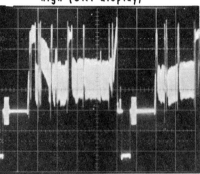

Photo 8.10   Left-hand edge
RV4 (CRT display)

Photo 8.11   Left-hand edge RV4
(monitor screen)

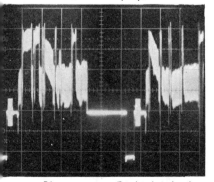

Photo 8.12   Right-hand edge
RV5 (CRT display)

Photo 8.13   Right-hand edge RV5
(monitor screen)

181

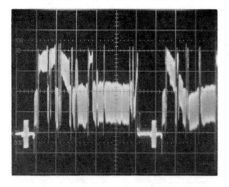

Photo 8.14   Composite sync
removed (CRT display)

Photo 8.15   Composite sync removed
(monitor screen)

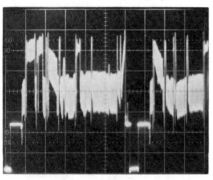

Photo 8.16   Colour burst
removed (CRT display)

Photo 8.17   Colour-burst removed
(monitor screen)

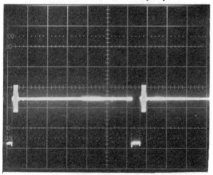

Photo 8.18   Video fade to
black (CRT display)

Photo 8.19   Picture distortion
(monitor screen)

Photo 8.20   Inverted picture (monitor screen)

● colour-burst defeat switch, S4. The picture should change to a black-andwhite one when S4 is closed; refer to Photos 8.16 and 8.17,

● video defeat switch, S5. Only the sync and colour burst information should be passed when S5 is closed, producing a blank screen, as can be seen from Photo 8.14.

Using the Video box

There are two major factors which can cause a reduction in performance from the Video Box. Firstly, if a +12 V d.c. power supply with poor smoothing or regulation is used, some of the functions will not work properly and the overall picture quality will suffer. For this reason, it is strongly recommended that if a ready-built d.c. supply is used, it should be a regulated 300 mA 12 V type (stock code YB23A). If you are constructing your own, it should incorporate a +12 V regulator, such as the

Video and TV projects

μA78M12UC (stock code QL29G). Please note that the Video Box is *not* reverse-polarity protected, so damage will occur to the circuit if the power supply connections are reversed. Secondly, if an inferior video signal is fed into the unit a poor quality signal will be produced. This can result in a dim, fuzzy and possibly unstable picture on the monitor screen.

The main function of the Video Box is to provide the ability for fading the picture information down to a blank screen. This effect can be used as a *fade-in* or *fade-out* on dubbed recordings (e.g. camcorder to VCR), lending a more professional appearance to your video productions. The whole picture, but not the sync, can be suddenly switched on and off by operating the video defeat switch, S5. The output defeat switch, S1, can appear to have a similar effect, but as it switches off *all* the signal components, your video equipment can take some time to resynchronise to the signal.

Two additional video effects have been provided for. When operated, the picture inverse switch, S3, produces a *negative* image. This unusual effect has often been exploited by pop video and sci-fi film makers. The colour burst defeat switch, S4, should have the effect of turning off the colour circuits within your video equipment. This function is commonly referred to as the *colour-killer* circuit (as discussed earlier), and when activated has the effect of producing a black-and-white picture. It is unlikely that you will ever need to use the sync defeat switch, S2, as this removes the composite sync from the video signal. However, if the Video Box is used as part of a larger video system, it may be necessary to switch off its own recovered sync, or possibly feed in external sync.

184

Terminal blocks

PL1	PL2	PL3	PL4	PL5	PL6	PL7
pin 1 = 0 V	pin 1 = 0 V	pin 1 = 10.8 V	pin 1 = 0 V	pin1 = 0 V	pin 1 = 0 V	pin 1 = 0 V
pin 2 = 0 V	pin 2 = 12 V	pin 2 = 12 V	pin 2 = 4.7 V	pin 2 = 0.5 V	pin 2 = 12 V	pin 2 = 0 V
	pin 3 = 0 V	pin 3 = 12 V				
	pin 4 = 0 V	pin 4 = 12 V				
	pin 5 = 12 V	pin 5 = 12 V				
	pin 6 = 12 V	pin 6 = 12 V				
	pin 7 = 0 V	pin 7 = 0 V				
	pin 8 = 12 V	pin 8 = 12 V				
	pin 9 = 0 V	pin 9 = 0 V				
	pin 10 = 0 V	pin 10 = 0 V				

Semiconductors

TR1	TR2	TR3	TR4	TR5	TR6	TR7	TR8	TR9	TR10	TR11
E = 6.3 V	E = 0 V	E = 3.2 V	E = 0.6 V	E = 3.9 V	E = 3 V	S = 3 V	E = 3 V	E = 11.5 V	E = 4 V	E = 3.2 V
B = 7 V	B = 0.1 V	B = 3.8 V	B = 1.3 V	B = 3.2 V	B = 3.6 V	G = 0.5 V	B = 3.6 V	B = 10.9 V	B = 4.7 V	B = 2.5 V
C = 12 V	C = 10.8 V	C = 12 V	C = 8.6 V	C = 3.6 V	C = 12 V	D = 3 V	C = 10.9 V	C = 2.5 V	C = 12 V	C = 0 V

IC1	IC2	IC3	IC4	IC5	IC6
pin 7 = 12 V	pin 8 = 12 V	pin 14 = 12 V	pin 16 = 12 V	pin 14 = 12 V	pin 14 = 12 V

Table 8.2 D.C. test measurements (read with a digital voltmeter)

Video box parts list

Resistors — All 0.6 W 1% metal film (unless specified)

R1	82 Ω	1	(M82R)
R2	6k8	1	(M6K8)
R3,29,32, 33,35,36, 39,41	10 k	8	(M10K)
R4,24, 42	470 Ω	3	(M470R)
R5,6,7,13, 14,16, 21, 22,28,43	1 k	10	(M1K)
R8,11, 27,40	2k7	4	(M2K7)
R9	560 Ω	1	(M560R)
R10	560 k	1	(M560K)
R12,34, 38,44	4k7	4	(M4K7)
R15,23	220 Ω	2	(M220R)
R17	1k5	1	(M1K5)
R18	390 Ω	1	(M390R)
R19	2k2	1	(M2K2,)
R20,25	100 Ω	2	(M100R)
R26	270 Ω	1	(M270R)
R30	150 Ω	1	(M150R)
R31	820 Ω	1	(M820R)
R37,46	15 k	2	(M15K)
R45	27 k	1	(M27K)
RV1,2	1 k hor encl preset	2	(UH00A)

| RV3 | 10 k min pot lin | 1 | (JM71N) |
| RV4,5,6 | 22 k hor encl preset | 3 | (UH04E) |

Capacitors

C1,9,14, 17,18,20, 21,27	100 nF minidisc	8	(YR75S)
C2,3,4, 10,11,15, 19,22	47 μF 25 V PC elect	8	(FF08J)
C5,13	22 pF polystyrene	2	(BX24B)
C6	560 pF polystyrene	1	(BX33L)
C7,8,24	100 nF polylayer	3	(WW41U)
C12	100 pF polystyrene	1	(BX28F)
C16	68 pF polystyrene	1	(BX27E)
C23	470 μF 16 V PC elect	1	(FF15R)
C25	680 pF polystyrene	1	(BX34M)
C26	2n2F polystyrene	1	(BX37S)
VC1	65 pF trimmer	1	(WL72P)

Semiconductors

IC1	LF351	1	(WQ30H)
IC2	LM1881N	1	(UL75S)
IC3	74HC4066	1	(UF10L)
IC4	4042BE	1	(QX19V)
IC5,6	4011BE	2	(QX05F)
TR1,2,3, 4,6,8,10	BC184L	7	(QB57M)
TR5,9,11	BC214L	3	(QB62S)
TR7	BF244A	1	(QF16S)
D1,2,4,5, 9,10,11	1N4148	7	(QL80B)

D3,6,7,8	BAR28	4	(QQ13P)
LD1	LED red	1	(WL27E)

Miscellaneous

PL1,4,5, 6,7	2-way minicon plug	5	(RK65V)
PL2,3	10-way minicon plug	2	(RK66W)
	2-way minicon hsng	5	(HB59P)
	10-way minicon hsng	2	(FY94C)
	minicon terminal	3	(YW25C)
	miniature coax	1	(XR88V)
	10-way ribbon cable	1	(XR06G)
	0.71 mm 2 swg TC wire	1	(BL14Q)
	8-pin DIL socket	2	(BL17T)
	14-pin DIL socket	3	(BL18U)
	16-pin DIL socket	1	(BL19V)
	PCB	1	(GE85G)
	instruction leaflet	1	(XT04E)
	constructors' guide	1	(XH79L)

Optional (not in kit)

RV3	10 k min slide pot lin	1	(JM85G)
	slide knob B	1	(YG09K)
	K14B knob	1	(FK39N)
	5 mm LED clip	1	(YY40T)
	SPST ultra min toggle	5	(FH97F)
	a.c. adaptor regulated	1	(YB23A)
	M6006 ABS console	1	(LH66W)
	75 Ω BNC round skt	2	(FE31J)
	2.5 pan mnt pwr skt	1	(JK10L)
	$1/4$ in no. 4 self tap screw	1	(FE68Y)

Take note — Take note — Take note — Take note

Possible modification to Video Box

It is reported that, when using the Video Box with certain VHS video recorders, a colour shift or distortion to the picture may be noticeable. To correct this, the following modification could be implemented to update your kit, or added while building it in the first place.

Add a 47 µH choke (WH39N) in series with resistor R21; see Figure 8.7. However, as there is no physical position on the PCB for this additional component, it must be fitted as shown in Figure 8.8. De-solder, and lift out of the board the lead of R21 nearest to the edge of the PCB. Next, insert the choke and solder it in place. Finally, solder the free end of R21 to the free end of the choke.

Video and TV projects

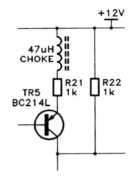

Figure 8.7   Modified circuit

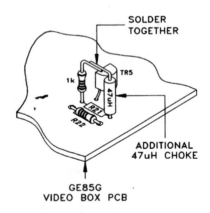

Figure 8.8   Positioning the choke